Mechanics of materials

An individualized approach

Mechanics of materials
An individualized approach

Edward Hornsey
University of Missouri-Rolla

David McFarland
Wichita State University

Karl Muhlbauer
University of Missouri-Rolla

Bert Smith
Wichita State University

Houghton Mifflin Company Boston
Atlanta Dallas Geneva, Illinois
Hopewell, New Jersey Palo Alto London

Printed in the U.S.A.
Library of Congress Catalog Card Number: 76–18470
ISBN: 0–395–24993–7

B 60022

Reference manual

Contents

Reference manual

Preface

This text consists of two distinctly different parts: The *Reference Manual* and the *Study Guide*.

The *Reference Manual* contains the theoretical foundations for engineering problems in mechanics of materials. This includes definitions, principles, laws, derivations of pertinent formulas, and homework problems and their answers. The *Study Guide* serves as a guide to the study of the material in the *Reference Manual*. The *Study Guide* is the controlling part of this learning package, and refers the learner to the *Reference Manual* when appropriate.

Emphasis should be placed on the importance of the material in the *Reference Manual*. Students frequently skim over theoretical material in their rush to begin solving problems. Consequently, important information, such as the assumptions which govern the derived equations, is often missed. This lack of emphasis frequently leads to incorrect results in the solutions of engineering problems.

It is our desire that the *Reference Manual* and *Study Guide* serve as an effective learning package for you now, and a useful reference source for you in the future.

E E H
D E M
K C M
B L S

Chapter one

Uniform stress and strain

R1-1 Definitions, stress–strain relationships, and Poisson's ratio

Definitions

The following definitions are presented now, so that we can proceed in a more orderly manner with the study of the topics which were introduced in Section 1-1 in the Study Guide.

Sign convention We shall adopt the standard sign convention for tension and compression, i.e., tension is positive and compression is negative. This convention applies to forces and stresses as well as to deformations and strains.

Normal stress Consider an area A which resists a force P acting perpendicular to the area A. The normal stress at any infinitesimally small portion of the area is defined as

$$\sigma = \lim_{\Delta A \to 0} \frac{\Delta P}{\Delta A} = \frac{dP}{dA} \tag{1}$$

where dP is the load acting on the infinitesimal area dA.

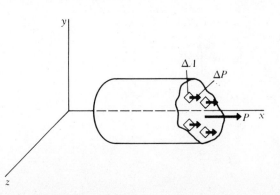

Uniform normal stress If we can assume the stress on an area to be of uniform (constant) intensity, then Equation (1) simplifies as follows:

$$dP = \sigma \, dA$$

$$\int dP = \int \sigma \, dA$$

$$P = \sigma A$$

or

$$\sigma = \frac{P}{A} \tag{2}$$

The stress given by Equation (2) is called the *uniform normal stress*. (When the stress varies over the area A, Equation (2) may still be used to obtain the average normal stress on the area.)

▶ **NOTE** For the normal stress to be of uniform intensity over the cross-sectional area, the following conditions must hold. The member on which the load acts must be straight, with a constant cross-sectional area. The load must be applied through the centroid of the cross-sectional area, and the member must be of *homogeneous material*. (*A material that has the same physical properties at every point is called a homogeneous material.*)

Normal strain Consider a fiber of material of length L, which can elongate (or contract) by an amount δ. The normal strain on any infinitesimally small segment of the fiber is defined as

$$\epsilon = \lim_{\Delta L \to 0} \frac{\Delta \delta}{\Delta L} = \frac{d\delta}{dL} \tag{3}$$

where $\Delta \delta$ is the elongation (or contraction) of the segment ΔL.

Uniform normal strain If we can assume the strain of a fiber of length L to be uniform (constant) along the length L, then Equation (3) simplifies as follows:

$$d\delta = \epsilon \, dL$$

$$\int d\delta = \int \epsilon \, dL$$

$$\delta = \epsilon L$$

or

$$\epsilon = \frac{\delta}{L} \tag{4}$$

The strain given by Equation (4) is called the *uniform normal strain*. (When the strain is not uniform, Equation (4) gives the average normal strain over the length L.)

▶ **NOTE** For the normal strain to be uniform along the complete length of the member, the same conditions must hold as given for the uniform normal stress defined by Equation (2).

Stress–strain diagram of low-carbon steels

The term *carbon steel* is used to designate those ferrous alloys in which carbon is the main element used to control the mechanical properties of the material alloy. The term *alloy steel* is used to designate those steels in which usually small, but precisely controlled, amounts of nickel, tungsten, vanadium, molybdenum, or chromium are used to control the physical and mechanical properties. Carbon steel is one of our most common structural materials. It is divided into four classes according to the carbon content.

Low-carbon steel contains from 0.08% to 0.35% carbon
Medium-carbon steel contains from 0.35% to 0.50% carbon
High-carbon steel contains more than 0.50% carbon
Cast steel contains about 2% carbon

Compositions containing more than 2% carbon are classified as *cast iron*.

Laboratory tests in tension or compression are usually performed on strictly standardized specimens at constant very slow rates of deformation and constant temperature. The tension test is most often used for relatively ductile (refer to Section R1-2) materials such as steel, aluminum, plastics, rubber, etc., while the compression test is usually used for brittle (refer to Section R1-2) materials such as concrete, brick, cast iron, glass, etc. Figure 1-1 on page 4 shows a stress–strain diagram for a typical low-carbon (or *mild*) steel, which was subjected to a uniaxial tension test.

More definitions

For a better understanding of the following definitions, you may need to reexamine the stress–strain diagram shown in Figure 1-1.

Proportional limit (PL) The highest stress on the initial straight-line portion of the stress–strain diagram.

Yield point (YP) The stress at which there is an appreciable increase in strain with no increase in stress, with the limitation that, if straining is continued, the stress will again increase. The yield point represents the transition from elastic to plastic behavior of the material.

Modulus of elasticity (E) The slope of the initial straight-line portion of the stress–strain diagram:

$$E = \frac{\sigma}{\epsilon} = \frac{\Delta\sigma}{\Delta\epsilon} \tag{5}$$

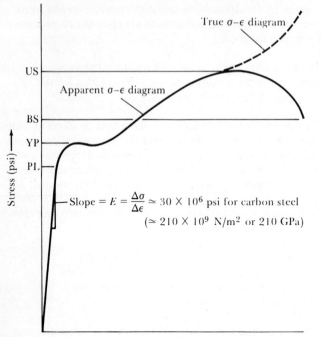

Figure 1-1

where

E = modulus of elasticity
σ = any stress *on the straight-line portion* of the stress–strain diagram
ϵ = strain corresponding to σ

▶ **NOTE** E is also known as *Young's modulus*, after Thomas Young, who in 1807 suggested the use of the stress-to-strain ratio to measure the stiffness of a material. The proportionality between load and deflection up to the proportional limit of a material is also known as *Hooke's law*, after Robert Hooke, who first observed this relationship and published it in 1678.

Elastic limit (EL) The greatest stress that a material is capable of withstanding without producing a permanent set of deformation upon release of that stress.

Ultimate strength (US) The highest stress which a material can withstand before rupture.

Breaking strength (BS) The stress at which the material ruptures.

▶ **NOTE** The stress–strain relationship shown by the solid lines in Figure 1-1 is often referred to as an *apparent* or *engineering stress–strain diagram*, because all stresses and strains are calculated with the *original* cross-sectional area and the *original* gage length. We know, however, that a material loaded in one direction will also undergo strains perpendicular to the direction of the load. You may easily demonstrate this fact by stretching a rubber band and observing the decrease of the cross-sectional area (the rubber band gets thinner) as the length increases.

When the force, area, and modulus of elasticity are constant over the length of the member, we can obtain a useful formula for the deformations of axially loaded members by substituting Equations (2) and (4) into Equation (5):

$$E = \frac{P/A}{\delta/L} = \frac{PL}{A\delta}$$

or

$$\delta = \frac{PL}{AE} \tag{5a}$$

When P, A, and E are constant for each of a finite number of sections n of an axially loaded member, we can use the summation

$$\delta = \sum_{i=1}^{n} \frac{PL}{AE} \tag{5b}$$

When P, A, and E change as functions of the length of the member, we can still obtain an approximate solution by assuming that the changing quantities are constant over an infinitesimal length dx. We then have the equation

$$\delta = \int_0^L \frac{P}{AE}\, dx \tag{5c}$$

where P, A, and E must be expressed as functions of the distance x along the member.

▶ **NOTE** Equations (5b) and (5c) give only approximate results, since one or more of the assumptions made in the derivation of Equations (2) and (4) do not hold. Nevertheless, these equations are quite useful, and are sufficiently accurate for most engineering applications.

Now let's look at some more definitions.

True stress–strain diagram When all the dimensional changes are taken into account in the calculation of stresses and strains, we obtain what is called a *true stress–strain diagram*. The general shape of a true stress–strain diagram for low-carbon steel is shown by the dashed lines in Figure 1-1. Note that the most pronounced deviation from the apparent stress–strain diagram occurs after the ultimate strength has been exceeded. The reason is that this steel will undergo a very pronounced reduction in cross-sectional area (called *necking*) at the section where rupture occurs. Thus the stress based on the actual cross-sectional area is much larger than that based on the original cross-sectional area.

The true stress–strain diagram is very seldom needed for engineering purposes. It is of little use in engineering design, since design is based on original dimensions.

Poisson's ratio The ratio of lateral strain to axial strain as obtained from a uniaxial tension test:

$$\mu = - \frac{\epsilon_\ell}{\epsilon_a} \tag{6}$$

where

μ = Poisson's ratio (a positive number)
ϵ_ℓ = lateral strain (measured perpendicular to the applied load)
ϵ_a = axial strain (measured in the direction of the applied load)

▶ **NOTE** Poisson's ratio is named after Simeon D. Poisson, who identified it in 1811. It is a constant for any *homogeneous, isotropic material.* (*A material the physical properties of which are the same in all directions at any given point is called isotropic.*) For strains within the elastic range of the material, its value lies between $\frac{1}{4}$ and $\frac{1}{3}$ for most metals. Beyond the elastic limit of the material, the value for μ is usually assumed to be about $\frac{1}{2}$.

The minus sign in Equation (6) makes Poisson's ratio a positive number because ϵ_ℓ and ϵ_a are always of opposite signs. For example, if ϵ_ℓ is an elongation (a positive quantity), then ϵ_a must be a contraction (a negative quantity).

Now return to the Study Guide for some examples of how these definitions may be used for the solution o˙ engineering problems.

■ **STOP**

R1-2 Stress-strain diagram of a structural aluminum alloy, definitions, introduction to design and analysis

Stress–strain diagram of a structural aluminum

Aluminum is very light (it weighs only about one-third as much as steel) and highly resistant to corrosion. In its pure state it is soft and of relatively low strength, but when alloyed with even small amounts of copper, silicon, manganese, or iron, its strength-to-weight ratio increases considerably. The most widely used structural aluminum alloys (also known as *duraluminum*) contain from 2.5% to 5.5% copper, and much smaller amounts of magnesium and manganese.

Aluminum alloys are widely used in aircraft and spacecraft construction, and are becoming ever more popular in the building industries.

Aluminum in its pure form is more corrosion resistant than its alloys. Thus, it is often used as a surfacing on duraluminum sheets to prevent corrosion. The resulting products are called *alclads*.

Figure 1-2 is the stress–strain diagram of a typical medium-strength structural aluminum alloy that has been subjected to a uniaxial tension test.

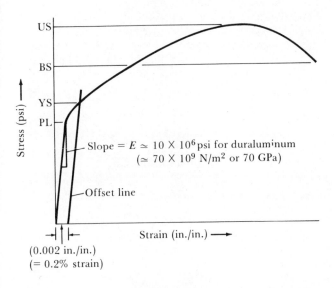

Figure 1-2

Compare Figure 1-2 with Figure 1-1. ■ **STOP**

A comparison of the stress–strain diagrams for mild steel and duraluminum shows that there is quite a difference between the two materials, especially in the shape of the two curves in the range between the proportional limit and ultimate strength. Specifically, duraluminum does not have a yield point. In order to have a quantity that corresponds to a yield point, it is customary to designate arbitrarily that stress which produces a certain predetermined amount of permanent deformation in the specimen as the *yield strength* of the material. This method of determining the yield strength is illustrated in Figure 1-2, where a line, offset by the arbitrary amount of 0.2% of strain, is drawn parallel to the straight-line portion of the stress–strain diagram. The stress at which this *offset line* intersects the curve is designated as the *yield strength of the material at 0.2% offset*. The 0.2% offset is most commonly used to determine the yield strength of any structural material that does not possess a well-defined yield point, but other values for the offset are sometimes specified for certain applications.

▶ **NOTE** In Section 1-1, you learned that the proportional limit, the elastic limit, and the yield point of a material are all defined differently. As an engineer you must know the exact definition of each of these terms, but for practical purposes the stress and strain values of these three points are often considered to be the same as the value of the yield point, or of the yield strength if the material does not have a yield point.

There are other important characteristic properties of materials that can be obtained from stress–strain diagrams. We shall make more extensive use of these properties later on in this text. It is sufficient for our purposes at this time simply to learn their definitions.

Ductility   The ability of a material to deform under load. The percent elongation of a specimen from zero stress to fracture and the corresponding percent reduction in cross-sectional area during a uniaxial tension test are relative measures of the ductility of a material:

$$\text{percent elongation} = \frac{L_f - L_i}{L_i} (100)\% \tag{7}$$

where

L_f = final gage length, L_i = initial gage length

$$\text{percent reduction in cross-sectional area} = \frac{A_i - A_f}{A_i} (100)\% \tag{8}$$

where

A_i = initial cross-sectional area

A_f = final cross-sectional area measured at point of fracture

Resilience   The capacity of a material to absorb energy in the elastic range. The *modulus of elastic resilience* MER is equal to the area under the straight-line portion of the stress–strain diagram.

$$\text{MER} = \frac{\sigma_{PL}}{2} \epsilon_{PL} \tag{9}$$

where

MER = modulus of elastic resilience
σ_{PL} = stress at the proportional limit
ϵ_{PL} = strain at the proportional limit

Substituting ϵ_{PL}/E and making the assumption that the proportional limit is almost equal to the yield point ($\sigma_{PL} \cong \sigma_{YP}$), we obtain

$$\text{MER} \cong \frac{1}{2} \frac{\sigma_{YP}^2}{E} \tag{10}$$

▶ **NOTE**   The area under the stress–strain curve represents the energy per unit volume absorbed by the material. Equation (10) shows that the material best suited to absorb high-impact stresses (a large amount

of energy) without undergoing permanent deformation should have a
high yield point and a low modulus of elasticity.

Plastic deformation The deformation that occurs beyond the elastic limit of the material. In general, a material that has the ability to undergo large plastic deformation before fracture is called a *ductile material*, while a *brittle material* is one that does not have this ability.

Toughness The capacity of a material to absorb energy during plastic deformation. The *modulus of toughness* MT is equal to the total area under the stress–strain diagram, and thus gives an indication of the maximum amount of energy that a material can absorb before fracture. A good approximation of the modulus of toughness may be obtained from the following equation:

$$MT = \frac{\sigma_{PL} + \sigma_{US}}{2} (\epsilon_{BS}) \tag{11}$$

where

$$MT = \text{modulus of toughness}$$

$$\frac{\sigma_{PL} + \sigma_{US}}{2} = \text{arithmetic average of proportional limit and ultimate strength}$$

$$\epsilon_{BS} = \text{strain at the breaking strength}$$

Equation (11) shows that materials that must be able to absorb high-impact loads without fracturing must have a high proportional limit, a high ultimate strength, and high ductility.

Introduction to design and analysis

As a practicing engineer you will most likely use mechanics of materials for two main purposes: design and analysis. We shall assume in this text that these two terms have the following meanings.

Design is the selection of the materials and the determination of the dimensions required for the proper functioning of a proposed structure or machine under a given set of environmental and loading conditions.

Analysis is the investigation of the stresses, deformations, and/or load-carrying capacity of a structure or machine that has already been designed.

Analysis requires only that the engineer know about the physical characteristics of the material, the type and magnitude of the applied loads, and the dimensions of each member. Design requires the collection and interpretation of much more data and general information concerning the intended use of the proposed structure or machine.

In the following sections we shall have more to say about design as it applies to the topics under consideration. At this time it suffices to discuss the terms *failure* and *factor of safety*, which are frequently used by designers. *Failure of a member occurs when it is no longer capable of performing the function for which it was designed.* Failure does not necessarily imply fracture; a member may cease to function properly for a wide variety of reasons. Some general forms of failure of a member are fracture, large elastic or inelastic deformations, buckling, and vibrations (the amplitude of vibration may become large enough to cause failure by collision with stationary parts, or to cause shaking, excessive noise, etc.). Remember, in engineering terminology, failure does not necessarily imply that something has been destroyed, but rather that it has ceased to function properly.

Factor of safety is the ratio of the load in a member which produces failure (failure load) to the load under which the member is expected to function (working load). We can also define factor of safety in terms of failure stress and working stress. Thus

$$\text{fs} = \frac{P_F}{P_W} \qquad \text{or} \qquad \text{fs} = \frac{\sigma_F}{\sigma_W} \tag{12}$$

where

fs = factor of safety
P_F = failure load, σ_F = failure stress
P_W = working load, σ_W = working stress

▶ **NOTE** The two definitions for the factor of safety yield the same results in the type of problems that we shall study in this text. Under certain conditions, which are beyond the scope of this text, the two definitions are not compatible. In general, you will always be correct if you base the factor of safety on the loads rather than on the stresses.

The proper selection of an appropriate factor of safety requires the careful consideration of at least the following items.

1 How accurate is the estimate of the loads which the member is expected to resist?
2 How reliable is the quality of the material to be used? (Will the actual member have the same physical characteristics as the test specimen on the basis of which the design calculations were made?)
3 All engineering design calculations are based on certain assumptions concerning the type of loads, stress distribution, etc. How valid are these assumptions?
4 Will the physical characteristics of the material change with time? (Corrosion, temperature fluctuations, and the duration and variation of loads must be considered under this item.)
5 How will the fabrication process affect the finished product? (Welding, for example, may alter the physical properties of a material, while improper alignment of rivet holes may affect the load distribution.)
6 Will the fabrication or manufacture of the structure or machine be

closely supervised, and will it be performed under conditions which allow close adherence to design specifications? (A structure which is erected under possibly adverse conditions in the field is obviously less likely to conform exactly to specifications than one which is assembled by skilled workmen in a sheltered environment.)

7 Will failure lead to loss of lives? What would be the economic consequences of failure?

Turn to the Study Guide for more examples and explanations of these concepts. ■ STOP

R1-3 Principle of superposition, the generalized Hooke's law for isotropic materials, shearing stresses and strains

The principle of superposition states that the total stresses and strains produced by several forces acting on a body is the algebraic sum of the stresses and strains that are produced by the forces acting separately. (There are some important limitations to this principle, which are stated under the next Caution sign.)

We may use the principle of superposition to determine strains in a body that is subjected to normal stresses in more than one direction. In Section SG1-1 of the Study Guide we defined the modulus of elasticity as

$$E = \frac{\sigma}{\epsilon}$$

and Poisson's ratio as

$$\mu = -\frac{\epsilon_\ell}{\epsilon_a}$$

where both definitions were based on the results of a uniaxial tension test. Figure 1-3 (page 12) illustrates how the principle of superposition may be used to determine the strains resulting from stresses applied in two directions simultaneously.

In Figure 1-3, we see that the application of σ_x produces a tensile strain in the direction of σ_x of magnitude

$$\epsilon_x = \frac{\sigma_x}{E}$$

and a compressive strain perpendicular to σ_x of magnitude

$$\epsilon_y = -\mu\epsilon_x = -\mu\frac{\sigma_x}{E}$$

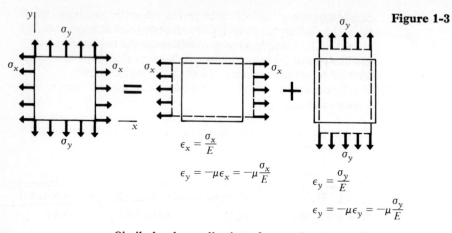

Figure 1-3

$$\epsilon_x = \frac{\sigma_x}{E}$$

$$\epsilon_y = -\mu\epsilon_x = -\mu\frac{\sigma_x}{E}$$

$$\epsilon_y = \frac{\sigma_y}{E}$$

$$\epsilon_y = -\mu\epsilon_y = -\mu\frac{\sigma_y}{E}$$

Similarly, the application of σ_y produces a tensile strain

$$\epsilon_y = \frac{\sigma_y}{E}$$

and a compressive strain

$$\epsilon_x = -\mu\epsilon_y = -\mu\frac{\sigma_y}{E}$$

Therefore, by the principle of superposition, we obtain the *total* strains in the x and y directions as follows:

$$\epsilon_x = \frac{\sigma_x}{E} - \mu\frac{\sigma_y}{E} \tag{13}$$

and

$$\epsilon_y = \frac{\sigma_y}{E} - \mu\frac{\sigma_x}{E} \tag{14}$$

Equations (13) and (14) are known as *the generalized Hooke's law* in two dimensions (biaxial stress). This law can easily be extended to cover the three-dimensional (triaxial) stress state by adding the strains produced by an additional stress in the z direction:

$$\epsilon_x = \frac{\sigma_x}{E} - \mu\frac{\sigma_y}{E} - \mu\frac{\sigma_z}{E} \tag{15}$$

$$\epsilon_y = \frac{\sigma_y}{E} - \mu\frac{\sigma_x}{E} - \mu\frac{\sigma_z}{E} \tag{16}$$

$$\epsilon_z = \frac{\sigma_z}{E} - \mu\frac{\sigma_x}{E} - \mu\frac{\sigma_y}{E} \tag{17}$$

Because of the cyclical nature of the terms, it is quite easy to memorize the generalized Hooke's law for triaxial stress, even though it looks rather nasty at first glance. At any rate, if you can't remember this law, you should always be able to develop it on your own. The use of these equations requires careful attention to our sign convention, i.e., tensile stresses and strains are positive and compressive stresses and strains are negative.

● **CAUTION** The generalized Hooke's law as stated above is valid only for homogeneous isotropic materials, i.e., materials having the same properties at each point and in all directions. Moreover, the *principle of superposition* is valid only if the strains are small and linearly related to the stresses that produce them, and if the material will obey Hooke's law under the combined action of all the stresses. Fortunately, many engineering structures satisfy all these requirements, and the principle of superposition is one of the engineer's most valuable tools.

Until now we have discussed only normal stresses, i.e., stresses produced by loads that are perpendicular to the area over which they act. Another very important category of stress is *shearing stress*, which is produced by forces that are parallel to the areas over which they act (see Figure 1-4). Shearing stress is designated by the lower-case Greek letter τ (tau).

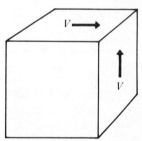 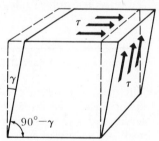

Figure 1-4

Shearing forces (V)   produce   Shearing stresses (τ) and
shearing strain (γ)

The expression for uniform shear stress is

$$\tau = \frac{V}{A} \tag{18}$$

where

V = shearing force
A = cross-sectional area over which V acts (this area is parallel to V)
τ = shearing stress (the stress is assumed to be uniformly distributed over the area A)

The determination of uniform shearing stress and the significance of such stress are discussed in more detail in Section 1-3 of the Study Guide, and in subsequent chapters.

Shearing stresses produce distortions of a member rather than axial deformations. These distortions are called *shearing strains,* and are designated by the lower-case Greek letter γ (gamma). (See Figure 1-4.)

We can obtain the relationship between shearing stress (τ) and shearing strain (γ) experimentally by twisting a cylindrical specimen in pure torsion and recording the distortion produced on the surface of the specimen. Since we shall not discuss torsion problems until Chapter 3, it is sufficient for our purposes here simply to note that, like a σ–ϵ diagram, the τ–γ diagram shows a definite proportional limit below which the shearing stress is directly proportional to the shearing strain. This relationship is described by the equation

$$G = \frac{\tau}{\gamma} \tag{19}$$

where

G = *modulus of rigidity* (the constant of proportionality also known as *shear modulus*)
τ = the shearing stress at or below the proportional limit
γ = the shearing strain corresponding to the shearing stress

Turn to the Study Guide. ■ **STOP**

R1-4 Thin-walled pressure vessels

Internal pressure developed in vessels such as propane tanks or boilers produces normal tensile stresses in the walls of such vessels. These tensile stresses vary in magnitude according to the thickness of the wall, but the variation is negligible if the wall thickness is small in comparison with the radius of the vessel. In this chapter we shall consider only spherical and cylindrical vessels, and we shall assume that the tensile stresses produced by the internal pressure are uniformly distributed over the entire wall. This assumption results in relatively simple expressions relating the internal pressure and wall thickness to the stress in the wall, and it is quite justified so long as the wall thickness is less than about one-tenth of the internal radius. (When the wall thickness is one-tenth of the internal radius, the difference between average stress in the thin-walled pressure vessel and the maximum stress obtained from the theory-of-elasticity solution is less than 5%. This much error is quite acceptable, because the pressure and the material properties are seldom known to a higher degree of accuracy.)

Figure 1-5(a) shows a thin-walled cylindrical pressure vessel of wall thickness t and average radius R (remember that for thin-walled vessels the difference between the inside radius and the outside radius is very small).

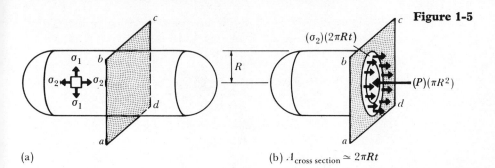

Figure 1-5

(a)

(b) $A_{\text{cross section}} \simeq 2\pi Rt$

When a vessel is subjected to an internal pressure P, normal tensile stresses of magnitude σ_1 and σ_2 are produced in the wall of the vessel, as shown. We sometimes refer to σ_1 as the *transverse* (or *circumferential*) *stress* and σ_2 as the *longitudinal* (or *axial*) *stress*, since they act in the circumferential and longitudinal directions, respectively.

Figure 1-5(b) is a free-body diagram obtained by passing the transverse section *abcd* through the vessel. Recall also (from fluid statics) that, in any direction, the component of the resultant force on any curved or inclined surface subjected to a uniform pressure is equal to the pressure multiplied by the projected area on a plane normal to that direction. From equilibrium considerations, i.e., from $\sum F = 0$, we find that

$$\sigma_2 (2\pi Rt) = P(\pi R^2)$$

or

$$\sigma_2 = \frac{PR}{2t} \tag{20}$$

where

σ_2 = longitudinal (or axial) stress (assumed to be uniformly distributed over the wall thickness)
P = internal pressure
R = average radius
t = wall thickness

Figure 1-6(a) shows the same vessel as Figure 1-5(a), and Figure 1-6(b) is the free-body diagram obtained by isolating that portion of the vessel identified by the section *efgh*. Again, from equilibrium considerations, we get

$$P(2RL) = 2\sigma_1(Lt)$$

or

$$\sigma_1 = \frac{PR}{t} \tag{21}$$

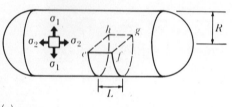

(a)

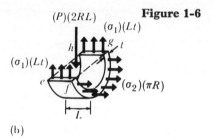

Figure 1-6

(b)

where

σ_1 = transverse stress (assumed to be uniformly distributed over the wall thickness)

P = internal pressure

R = average radius

t = wall thickness

L = an arbitrary length *ef* of the section *efgh*

▶ **NOTE** Comparing Equations (20) and (21), we see that the transverse stress (σ_1), that is, the stress in the short direction of a cylindrical pressure vessel, is always twice as large as the longitudinal stress (σ_2). Because of the symmetry of a spherical vessel, diametral sections similar to *abcd* in Figure 1-5(a) can be cut in any direction. The stresses in a spherical vessel are therefore equal in all directions, and their magnitude is given by Equation (20).

We may apply the equations derived in this section directly. They will give reliable results for the analysis and the design of vessels whose shape and dimensions are as specified in the opening paragraph, if they are constructed of homogeneous, isotropic material. By not limiting ourselves to the final form of these equations, but still following the steps taken in their derivation, we can solve a wide variety of problems, as will be shown in the Study Guide.

● **CAUTION** One important limitation of the method described above is the assumption that the normal stresses are uniformly distributed over the cross section of the wall of the pressure vessel. It has already been mentioned that this assumption is not correct for thick-walled vessels. The assumption is also inaccurate for areas on the vessel in which abrupt changes of shape occur, such as at the flat capped ends of a boiler or around valve connections or pipe inlets, etc. Such areas must be designed and analyzed by special methods.

Turn to the Study Guide. ■ **STOP**

Problems

▶ **NOTE** In all the following problems, the weights of the individual members of the structures are considered negligible.

1-2.1 A 1000-lb homogeneous body with its center of gravity at G is supported by a $\frac{1}{4}$-in.-diameter cable at D, a roller at A, and a $\frac{1}{4}$ in. × $\frac{1}{8}$-in. rectangular link at B. Determine the stresses in the cable and in the link.

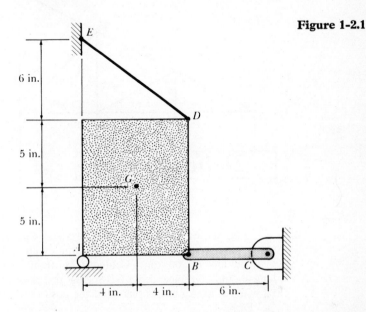

Figure 1-2.1

1-2.2 A 500-kg mass is supported by a horizontal beam, as shown. The stress in the cable AB is not to exceed 140 MPa. Determine the smallest permissible diameter of the cable. (Assume that $g = 9.8$ m/s².)

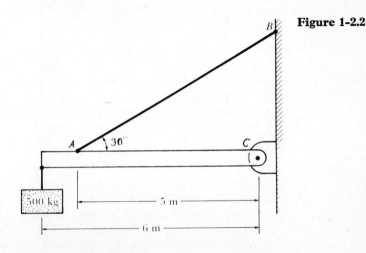

Figure 1-2.2

1-2.3 A shop sign is suspended from a truss, as shown. The sign weighs 2000 lb; one-half of this weight may be assumed to act at joint A and the other half at joint D. All members of the truss are made of structural steel which has a yield point of 58,000 psi. Determine the required cross-sectional areas of members CB, DE, and CE. Use a factor of safety of 2.

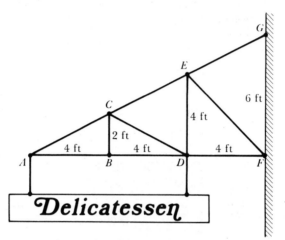

Figure 1-2.3

1-2.4 A 150-kg platform is supported by two steel cables, as shown. Cable AB has a cross-sectional area of 100 mm^2, while cable CD has a cross-sectional area of 200 mm^2. Cable AB is originally 10 m long. How long should you make cable CD so that the platform will be horizontal when a 900-kg box is placed in the position shown? ($E_{st} = 210 \times 10^9$ Pa.)

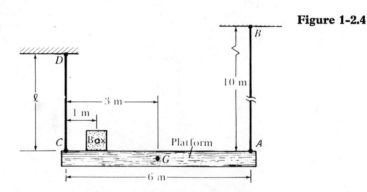

Figure 1-2.4

1-2.5 A weight W is suspended from three cables, as shown. The cross-sectional area of cable AB is 0.14 in^2 and that of cable BC is 0.10 in^2. Assuming the same material and factor of safety as in Problem 1-2.3, determine the maximum allowable load W. Is this a well-designed support structure? What is the required cross-sectional area of BD?

Figure 1-2.5

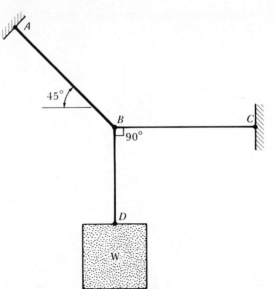

1-2.6 A triangular plate of negligible weight is suspended from three wires, as shown.

a) Where must a 3000-lb weight be placed on the plate so that each wire will carry the same load?

b) Wire A has a cross-sectional area of 0.1 in^2. What must be the cross-sectional areas of wires B and C so that the plate will remain horizontal when the weight is positioned as determined in part (a)? All wires are of the same material.

Figure 1-2.6

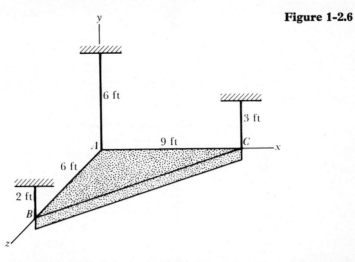

1-3.1 An aluminum bar and a steel bar are securely fastened together. What uniformly distributed load P (acting on the aluminum bar) is required to cancel the deformation of end A produced by the 40,000-lb force? (In other words: Find the load P which will make the relative deflection between A and C equal to zero.) ($E_{st} = 30 \times 10^6$ psi, $E_{al} = 10 \times 10^6$ psi)

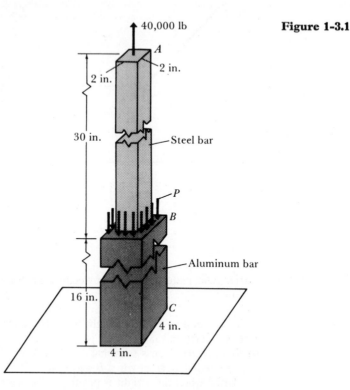

Figure 1-3.1

40,000 lb

A

2 in.

2 in.

30 in.

Steel bar

P

B

Aluminum bar

16 in.

C

4 in.

4 in.

1-3.2 Compute the change in the lateral dimensions of the steel and aluminum bars of Problem 1-3.1 when the 40,000-lb force and the 140,000-lb load are applied. ($\mu_{st} = \tfrac{1}{3}$, $\mu_{al} = \tfrac{1}{4}$)

1-3.3 The steel bar shown in Figure 1-3.3 is suspended at end B, has constant cross-sectional dimensions over its length of 15 m, and its total

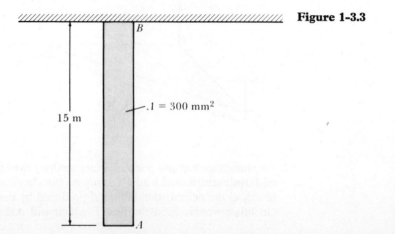

Figure 1-3.3

B

$A = 300$ mm^2

15 m

A

mass is 35 kg. Determine the displacement of end A with respect to end B. ($E_{st} = 200$ GPa)

1-3.4 A timber beam is supported by a 15-in.-long steel wire at end A and a 20-in.-long timber post at end B.

a) Where should the 10,000-lb load be applied so that both supports deform elastically by the same amount? ($E_{st} = 30 \times 10^6$ psi, $E_{wood} = 2 \times 10^6$ psi)
b) Calculate the area of the bearing plate at A, given that the allowable bearing stress is 500 psi for the wood. Also determine the dimension a of the square bearing plate at B, given that the allowable bearing stress on the soil is 33.3 psi. (The supports at A and D can be assumed to produce only vertical reactions.)

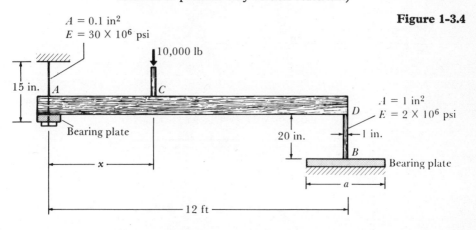

Figure 1-3.4

1-3.5 The allowable shear stress for the steel bolt at B is 70 MPa, and the ultimate strength of the wood block in shear parallel to the grain is 3.5 MPa. All members of the structure have widths of 25 mm. Determine the following (see Figure 1-3.5 on page 22):

a) The diameter required for the steel bolt.
b) The required distances d and s in the wood block. Use a factor of safety of 4 for the wood.

1-3.6 Member AD is made of wood ($E = 20$ GPa, $A = 323$ mm²), and member BC is made of aluminum ($E = 70$ GPa, $A = 65$ mm²). Calculate the strain in members AD and BC when a 27-kN force is applied as shown. (See Figure 1-3.6 on page 22.)

1-3.7 All the pins in the pin-connected structure shown have a cross-sectional area of 0.25 in² and act in single shear. Determine the shear stress in the pins A, E, H, and D. Also determine the required cross-sectional area of the cable, assuming that it is made of a steel whose ultimate strength is 100,000 psi. Use a factor of safety of 4; see Figure 1-3.7 on page 23.

1-4.1 A cylindrical pressure vessel of 1.3 m diameter is to operate in a vertical position, as shown. The water may reach a maximum level of

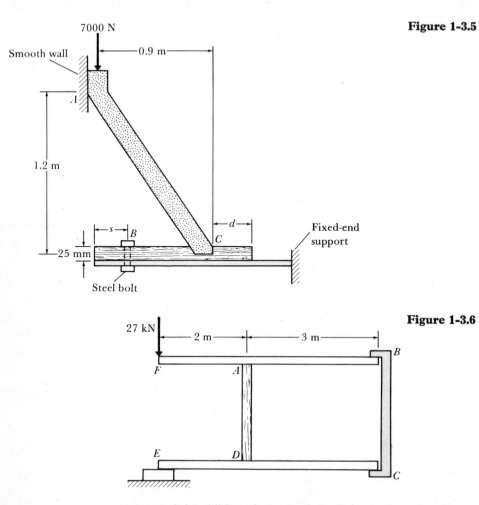

Figure 1-3.5

Figure 1-3.6

11 m and, in addition, the space above the water may then be pressurized to 240 kPa. Assuming that the allowable tensile stress is 70 MPa, calculate the wall thickness for a factor of safety of 2. (The mass of water is 1000 kg/m³.)

1-4.2 A cylindrical compressed-air tank 4 ft in diameter is 24 ft long and subjected to an internal pressure of 200 psi. What must the minimum wall thickness be if the maximum normal stress is not to exceed 20,000 psi and the total elongation over the middle 20 ft is not to exceed 0.0385 in? ($E = 30 \times 10^6$ psi, $\mu = \frac{1}{4}$)

1-4.3 A large cylindrical pressure vessel, 3 m in diameter and 11 m long, is used for processing rubber. It operates at a pressure of 830 kPa. If the wall thickness is 25 mm, what will be the increase in the diameter's dimension caused by the operating pressure? ($E = 200$ GPa, $\mu = 0.3$)

1-4.4 A boiler made of 12-mm-thick steel plate is 1 m in diameter, 2.5 m long, and operates at a pressure of 3.5 MPa. Calculate the change in

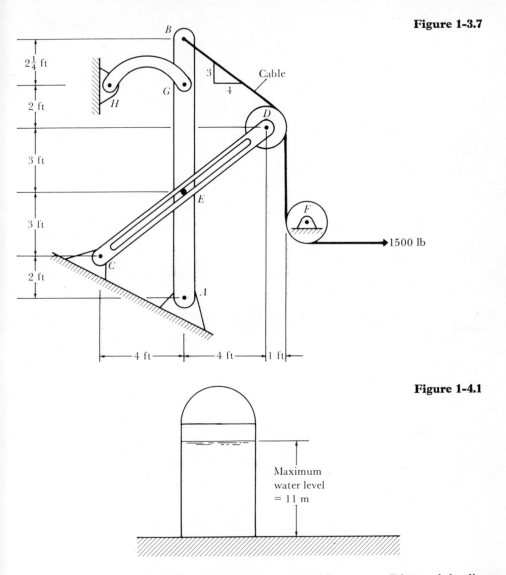

Figure 1-3.7

Figure 1-4.1

thickness of the steel plate due to this pressure. Disregard the direct compressive effect of the pressure on the inside wall. ($E = 200$ GPa, $\mu = \frac{1}{4}$).

1-4.5 Two hemispherical shells of 10-in. radius and 0.1-in. wall thickness are bolted together by four $\frac{1}{2}$-in.-diameter bolts. What is the elongation of the vessel over a 2-in. gage length when the stress in the bolts is 12,000 psi? ($E = 15 \times 10^6$ psi, $\mu = \frac{1}{4}$)

1-4.6 A $\frac{1}{4}$ in. × 1 in. steel ring is to be used to hold together the two halves of a solid cylinder, as shown in the figure. The inside diameter of the ring is $\frac{1}{25}$ in. less than the outside 5-ft diameter of the cylinder. The ring is heated until it can be slipped over the cylinder and is then allowed to cool. Assuming that the ring behaves elastically and that

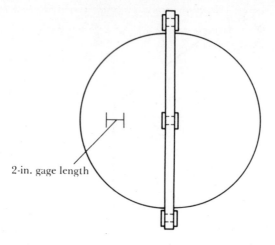

Figure 1-4.5

2-in. gage length

the cylinder is rigid (that is, its diameter is not changed by the pressure produced by the ring), calculate:

a) The tensile stress in the ring after it has cooled to room temperature,

b) The pressure of the ring on the cylinder.

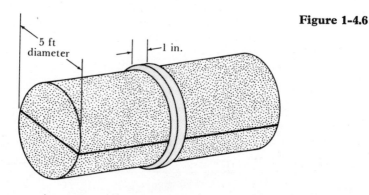

Figure 1-4.6

5 ft diameter

1 in.

1-4.7 You are asked to design a small tank which is to be filled with compressed air at a maximum pressure of 175 psi. Two shapes for the tank are under consideration:

a) Spherical with a 12-in. diameter,

b) Cylindrical, closed by hemispherical ends, with a 6-in. diameter. Specify the wall thickness for each. Assume a yield point of 30,000 psi and apply a factor of safety of 4. What must be the overall length of the cylinder to provide the same volume as the sphere?

Chapter two

Stress and strain transformation

R2-1 Stress at a point, plane stress, the wedge method

Stress at a point

The expression *stress at a point* refers to the state of stress on an *infinitesimal rectangular parallelepiped* (or simply a *rectangular element*).

We usually give the state of stress on each face of a rectangular element in terms of orthogonal components: one normal-stress and two shearing-stress components. A six-sided rectangular element may thus be subjected to eighteen stresses. However, these stresses are not all independent. Since finite stresses can change only by infinitesimal amounts over infinitesimal distances, the stresses on parallel faces of our element are assumed to occur in pairs of equal magnitude but opposite directions. The general state of stress on the three visible faces of a rectangular parallelepiped is shown in Figure 2-1.

Figure 2-1

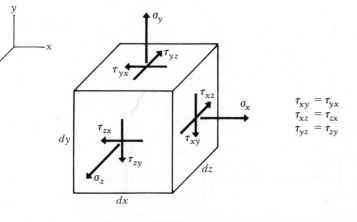

$$\tau_{xy} = \tau_{yx}$$
$$\tau_{xz} = \tau_{zx}$$
$$\tau_{yz} = \tau_{zy}$$

Each stress in Figure 2-1 is shown with a single arrow, but actually represents uniformly distributed stress over its respective area. *Unless specifically stated otherwise, a single arrow will be used throughout this book to represent uniformly distributed stress.*

Figure 2-1 shows nine stresses. Only six of these stresses are independent, since shearing stresses occur in pairs of equal magnitude on orthogonal faces of an infinitesimal element (that is, $\tau_{xy} = \tau_{yx}$, $\tau_{xz} = \tau_{zx}$, and $\tau_{yz} = \tau_{zy}$). In our discussion of plane stress, we shall prove that $\tau_{xy} = \tau_{yx}$.

The two-letter subscript in the notation for shearing stress gives the plane and the direction of the shearing stress. The first subscript letter denotes the normal to the plane on which the shearing stress acts, and the second subscript letter denotes its direction (for example, τ_{xy} acts on a plane perpendicular to the x axis and it is parallel to the y axis). A discussion of sign conventions will be deferred to Section R2-2.

In summary, six stresses completely define the state of stress at a point in a loaded body. They are usually given as three normal stresses and three shearing stresses in an orthogonal coordinate frame.

Plane stress

We shall be considering primarily that state of stress known as *plane stress*. A plane stress in the xy plane is defined by $\sigma_z = \tau_{xz} = \tau_{yz} = 0$, and is usually shown on a two-dimensional figure for simplicity. Figure 2-2(a) shows a state of plane stress on a rectangular parallelepiped, and Figure 2-2(b) gives the two-dimensional representation of the same stress state. It is important to realize that the stresses shown in the two-dimensional view (Figure 2-2b) act on planes perpendicular to the plane defined by $ABCD$. Also all stresses are assumed to be uniformly distributed over the planes on which they act. We shall usually represent a plane-stress state in a two-dimensional sketch. However, keep in mind that in actuality we are always dealing with three-dimensional elements.

If we multiply the uniform stresses of Figure 2-2 by the respective areas over which they act, we obtain forces. We can then show these

Figure 2-2

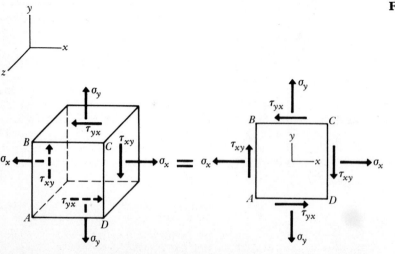

(a)   (b)

forces on a free-body diagram, to which the static equations of equilibrium apply.

● **CAUTION** Although having magnitude and direction, stresses do not obey the parallelogram law of vector addition, and are therefore not vectors. The static equilibrium equations can*not* be applied directly to stresses. We must first multiply the stresses by their respective areas to obtain forces. Only then can we apply the static-equilibrium equations.

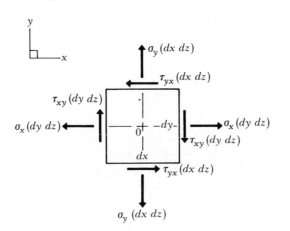

Figure 2-3

In Figure 2-3 we have a free-body diagram showing the forces that result from a plane-stress state. Since the stresses are assumed to be uniformly distributed, the resulting normal forces act through the center of their respective faces. Summation of moments about the center O of the element gives only the moments of the two shearing-force couples, since all normal forces intersect at O.

$$\sum^{+\curvearrowleft} M_O = \tau_{yx}(dx\ dz)\ dy - \tau_{xy}(dy\ dz)\ dx = 0 \qquad (1)$$

Therefore

$$\tau_{yx} = \tau_{xy} \qquad (2)$$

We have shown that the shearing stresses are equal in magnitude and that, to satisfy the conditions for equilibrium of moments, one of the shearing couples must be clockwise and the other counterclockwise. We could also have examined the free-body diagram for the general state of stress and shown that the other shearing-stress pairs must also be equal to satisfy the conditions for equilibrium of moments (that is, $\tau_{yz} = \tau_{zy}$ and $\tau_{xz} = \tau_{zx}$).

The state of plane stress is defined completely by three stresses: two normal stresses and one shearing stress. You must remember that one shearing stress cannot exist by itself. On an infinitesimal rectangular element in a plane-stress state, there must be four shearing stresses of equal magnitude. If the element is to be in equilibrium, they must form opposing shearing couples.

The wedge method

If we are given the stresses on any two planes of an infinitesimal element that is in a state of plane stress, we can determine the normal and shearing stresses on any specified third plane of the resulting wedge-shaped element. For example, if we are given σ_x, σ_y, and τ_{xy}, we can determine σ_θ and τ_θ by means of a free-body diagram, as shown in Figure 2-4. We simply write two equilibrium equations and solve for the two unknowns. In Figure 2-4, the area of the inclined plane is assumed to be dA. The areas of the vertical and horizontal planes are then given in terms of dA and the angle θ.

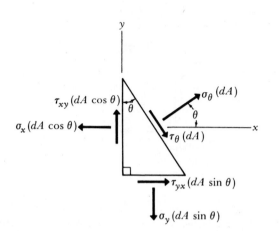

Figure 2-4

The method outlined in the preceding paragraph is known as the *wedge method* for determining stresses. The wedge method involves the following steps.

a) Sketch a free-body diagram of the wedge and include the shearing and normal stresses on each plane.
b) Multiply the stresses by the respective areas over which they act, to obtain forces.
c) Apply the static-equilibrium equations to solve for the unknowns.

You can accomplish step (c) by summing the forces in the horizontal and vertical directions, or you may find it advantageous to sum the forces normal and tangent to the inclined plane. The choice of equilibrium equations will depend on which quantities are unknown. If σ_θ and τ_θ are the unknowns, it is generally more direct to sum the forces parallel and perpendicular to the inclined plane, thus avoiding the coupled simultaneous equations that would result from summing forces in the horizontal and vertical directions.

As you can see, the wedge method is simply a problem in statics. Other variations of this problem are possible. The wedge need not include a right angle, and the unknowns need not be σ_θ and τ_θ. The only requirement is that the number of unknowns not exceed the number of independent static-equilibrium equations. A two-dimensional free-body diagram will yield at most three independent equilibrium equations. When two sides of the wedge are at right angles, one of the equilibrium equations shows that $\tau_{xy} = \tau_{yx}$, so that only two independent equations remain to be solved for unknowns.

Turn to the study program in Section SG2-1 of the Study Guide.
■ STOP

R2-2   Stress-transformation equations

The wedge method of analyzing plane stress involves the same mathematical procedures each time we employ it. Therefore it would be to our advantage to perform a wedge-method analysis of the most general state of plane stress, expressing both the known and unknown stresses in symbolic form and letting the orientation of the inclined plane be at some arbitrary angle θ. The resulting equilibrium equations are called the *transformation equations*.

Consider the problem in which σ_x, σ_y, and τ_{xy} are known and we are required to find the stresses on a plane oriented at angle θ counterclockwise from the vertical, as shown in Figure 2-5. The transformation equations are those which relate the stresses associated with the *nt* coordinates to the stresses associated with the reference *xy* coordinates.

Figure 2-5

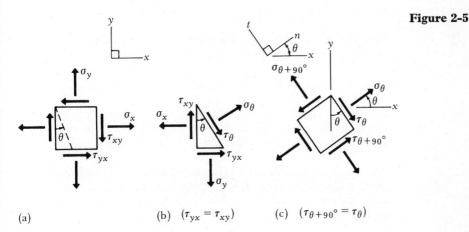

(a) (b) $(\tau_{yx} = \tau_{xy})$ (c) $(\tau_{\theta+90°} = \tau_\theta)$

▶ **NOTE** All three elements in Figure 2-5 represent the state of stress at the same point in the loaded body, but on different planes through that point.

If the area of the inclined plane in Figure 2-5(b) is dA, then the area of the vertical plane is $dA \cos \theta$, and the area of the horizontal plane is $dA \sin \theta$. A free-body diagram showing the forces can be constructed as shown in Figure 2-6. We can solve for the unknown normal stress σ_θ by summing the forces perpendicular to the inclined plane.

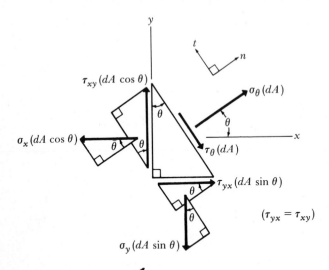

Figure 2-6

$$\sum^{+\nearrow} F_n = \sigma_\theta \, dA + \tau_{xy}(dA \cos \theta) \sin \theta - \sigma_x(dA \cos \theta) \cos \theta$$
$$+ \tau_{yx}(dA \sin \theta) \cos \theta - \sigma_y(dA \sin \theta) \sin \theta = 0. \qquad (3)$$

After solving for σ_θ in Equation (3) and making the substitution $\tau_{yx} = \tau_{xy}$, we obtain Equation (4):

$$\sigma_\theta = \sigma_x \cos^2 \theta + \sigma_y \sin^2 \theta - 2\tau_{xy} \sin \theta \cos \theta \qquad (4)$$

Similarly, we can determine the equation for τ_θ by summing the forces parallel to the inclined plane. The resulting equation is

$$\tau_\theta = (\sigma_x - \sigma_y) \sin \theta \cos \theta + \tau_{xy}(\cos^2 \theta - \sin^2 \theta) \qquad (5)$$

We could have summed forces horizontally and vertically to obtain Equations (4) and (5), but that would have required us to solve two coupled simultaneous equations.

The transformation equations (Equations 4 and 5) are frequently given in alternative forms involving the double angles. We can obtain these alternative forms by employing the following trigonometric identities:

$$\sin^2 \theta = \frac{1 - \cos 2\theta}{2}, \qquad \cos^2 \theta = \frac{1 + \cos 2\theta}{2}$$

$$2 \sin \theta \cos \theta = \sin 2\theta$$

After substituting these trigonometric identities into Equations (4) and (5), and simplifying, we obtain the results

$$\sigma_\theta = \frac{\sigma_x + \sigma_y}{2} + \left(\frac{\sigma_x - \sigma_y}{2}\right) \cos 2\theta - \tau_{xy} \sin 2\theta \tag{6}$$

and

$$\tau_\theta = \left(\frac{\sigma_x - \sigma_y}{2}\right) \sin 2\theta + \tau_{xy} \cos 2\theta \tag{7}$$

Although we have not formally stated our sign convention in the derivation of the transformation equations [Equations (4) and (5) or the double-angle form, Equations (6) and (7)], that convention was actually established when we sketched the stresses on the elements in Figure 2-5. The stresses and angle shown in Figure 2-5 establish the positive conventions. Therefore we can state our sign conventions as follows.

a) Tensile normal stresses (σ_x, σ_y, σ_θ) are positive and compressive normal stresses are negative.
b) The shearing stresses τ_{xy} and τ_θ are positive if they cause a clockwise moment on the element, and negative if they cause a counterclockwise moment on the element.
c) A positive angle θ is measured counterclockwise from the x axis to the action line of σ_θ, or from the y axis to the plane on which σ_θ acts. (Clockwise angles are negative.)

Turn to Section SG2-2 and continue with the study program.

■ STOP

R2-3 Principal stresses, maximum shearing stress

The wedge method and the resulting stress transformation equations are sufficient for determining stresses on inclined planes, provided that we know the orientations of the planes on which the stresses of interest occur. Such knowledge is available in the cases of the plane of the grain of wood (as we saw in Example 3 of Section R2-2), the plane formed by a weld, fracture planes, bedding planes in rock, etc. Frequently, however, we are interested in determining the maximum and minimum shearing and normal stresses and their orientations relative to our *xy* reference frame. We can derive equations for maximum and minimum stresses and their associated orientations by applying the principles we learned in differential calculus to the plane-stress transformation equations of Section R2-2.

Principal stresses

The normal-stress transformation equation is

$$\sigma_\theta = \frac{\sigma_x + \sigma_y}{2} + \left(\frac{\sigma_x - \sigma_y}{2}\right) \cos 2\theta - \tau_{xy} \sin 2\theta \tag{8}$$

To obtain the angles that give maximum and minimum values of σ_θ, we must set the derivative $d\sigma_\theta/d\theta$ equal to zero. The resulting equation is

$$\frac{d\sigma_\theta}{d\theta} = - \left(\frac{\sigma_x - \sigma_y}{2}\right) 2 \sin 2\theta_p - 2\tau_{xy} \cos 2\theta_p = 0$$

from which we obtain

$$\frac{\sin 2\theta_p}{\cos 2\theta_p} = \tan 2\theta_p = \frac{-\tau_{xy}}{(\sigma_x - \sigma_y)/2} \tag{9}$$

The angles θ_p are the angles from the x axis to the lines of action of the maximum and minimum normal stresses (or the angles from the y axis to the planes on which the maximum and minimum normal stresses act). There are two angles θ_p which satisfy Equation (9), and they are always 90° apart (for example, if $\tan 2\theta_p = 1.73$, then $2\theta_p = 60°$ and $180° + 60°$, and $\theta_p = 30°$ and $120°$).

The two angles determined by Equation (9) will be designated as θ_{p1} and θ_{p2}. When we substitute these angles into Equation (8), we obtain the maximum and minimum normal stresses:

$$\sigma_{p1} = \frac{\sigma_x + \sigma_y}{2} + \left(\frac{\sigma_x - \sigma_y}{2}\right) \cos 2\theta_{p1} - \tau_{xy} \sin 2\theta_{p1} \tag{10a}$$

$$\sigma_{p2} = \frac{\sigma_x + \sigma_y}{2} + \left(\frac{\sigma_x - \sigma_y}{2}\right) \cos 2\theta_{p2} - \tau_{xy} \sin 2\theta_{p2} \tag{10b}$$

We shall refer to the algebraically larger stress as σ_{p1} (the maximum normal stress), and the algebraically smaller stress as σ_{p2} (the minimum normal stress). These maximum and minimum normal stresses are called *principal stresses*, and their associated planes are called *principal planes*.

Equations (10) give the principal stresses in terms of the principal angles. We can also express the principal stresses directly in terms of σ_x, σ_y, and τ_{xy}, since the principal angles are functions of these stresses. Knowing the value of $\tan 2\theta_p$ as determined by Equation (9), we can construct the right triangle

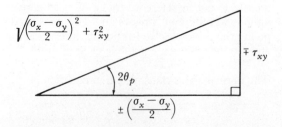

From this triangle we obtain:

$$\sin 2\theta_p = \frac{\mp \tau_{xy}}{\sqrt{[(\sigma_x - \sigma_y)/2]^2 + \tau_{xy}^2}}$$

$$\cos 2\theta_p = \frac{\pm(\sigma_x - \sigma_y)/2}{\sqrt{[(\sigma_x - \sigma_y)/2]^2 + \tau_{xy}^2}}$$

We can obtain the maximum and minimum normal stresses by substituting $\sin 2\theta_p$ and $\cos 2\theta_p$ into Equation (8). After simplifying, we get

$$\sigma_{p1,p2} = \frac{\sigma_x + \sigma_y}{2} \pm \sqrt{\left(\frac{\sigma_x - \sigma_y}{2}\right)^2 + \tau_{xy}^2} \qquad (11)$$

▶ **NOTE** Equations (10) and (11) both give principal stresses. Equation (11) has the advantage of enabling us to evaluate the principal stresses independently of the calculation for principal angles. However, it has the disadvantage of not indicating immediately which principal angle is associated with which principal stress, whereas when we evaluate principal stresses by means of Equations (10), we know which angles are associated with which principal stresses. On the other hand, in the latter case, there is a corresponding increase in the possibility of error: That is, an error in evaluating the principal angles will be carried through the principal stress calculations. Also experience indicates that errors are frequently made in evaluating $\cos 2\theta_p$ and $\sin 2\theta_p$. Perhaps the safest approach (to avoid mistakes) is to evaluate principal stresses by means of Equation (11), principal angles by means of Equation (9), and then use Equations (10) to check the principal stresses and associate the angles with the corresponding principal stresses. Additional methods for associating the principal stresses and principal angles are discussed in Sections SG2-3 and SG2-4 of the Study Guide, and in Section R2-4 of the Reference Manual.

When we add the two principal stresses of Equation (11), we obtain the relationship

$$\sigma_{p1} + \sigma_{p2} = \sigma_x + \sigma_y \qquad (12a)$$

Substituting θ and $\theta + 90°$ into Equation (8) and adding the two resulting equations yield

$$\sigma_\theta + \sigma_{\theta+90°} = \sigma_x + \sigma_y \qquad (12b)$$

▶ **NOTE** Equations (12) demonstrate an invariant property of normal stresses: Under a rotation of coordinates, the sum of the normal stresses is a constant.

Another important fact about principal stresses is that they always occur on planes of zero shearing stress. We can demonstrate this fact by substituting the $\sin 2\theta_p$ and $\cos 2\theta_p$ into the shear-stress-transformation equation as follows:

$$\tau_\theta = \left(\frac{\sigma_x - \sigma_y}{2}\right) \sin 2\theta + \tau_{xy} \cos 2\theta, \tag{13}$$

$$\tau_{\theta_p} = \left(\frac{\sigma_x - \sigma_y}{2}\right) \left[\frac{\mp \tau_{xy}}{\sqrt{[(\sigma_x - \sigma_y)/2]^2 + \tau_{xy}^2}}\right]$$

$$+ \tau_{xy} \left[\frac{\pm (\sigma_x - \sigma_y)/2}{\sqrt{[(\sigma_x - \sigma_y)/2]^2 + \tau_{xy}^2}}\right]$$

Therefore

$$\tau_{\theta_p} = 0 \tag{14}$$

Maximum shearing stress

We can obtain the maximum shearing stress in the same manner as the principal stresses. If we determine $d\tau_\theta/d\theta$ by means of Equation (13) and set it equal to zero, we obtain

$$\tan 2\theta_q = \frac{\sigma_x - \sigma_y}{2\tau_{xy}} \tag{15}$$

There are two angles θ_q that will satisfy Equation (15). These two angles locate the planes (positive angles are counterclockwise from the y axis) of maximum and minimum shearing stress. As in the case of the principal angles θ_p, the double angles $2\theta_q$ are 180° apart, so that the angles θ_q are 90° apart.

▶ **NOTE** The tangent of the double angles for maximum and minimum shearing stress (Equation 15) is the negative reciprocal of that obtained for principal stresses (Equation 9). Therefore the double-angle solutions for principal stresses ($2\theta_p$) are at right angles to the double angles ($2\theta_q$) associated with the planes of maximum and minimum shearing stress. The angles θ_p are at 45° angles to the angles θ_q.

The two angles determined by Equation (15) will be designated as θ_{q1} and θ_{q2}. If we substitute these angles into Equation (13), we obtain the maximum and minimum values of shearing stress:

$$\tau_{\theta_{q1}} = \left(\frac{\sigma_x - \sigma_y}{2}\right) \sin 2\theta_{q1} + \tau_{xy} \cos 2\theta_{q1} \tag{16a}$$

$$\tau_{\theta_{q2}} = \left(\frac{\sigma_x - \sigma_y}{2}\right) \sin 2\theta_{q2} + \tau_{xy} \cos 2\theta_{q2} \tag{16b}$$

▶ **NOTE** Remember that, on any infinitesimal rectangular element, a clockwise shearing couple on one set of faces must be accompanied by a counterclockwise shearing couple on the other two faces. Also we previously established the sign convention that shearing stresses τ_θ forming a clockwise couple are positive, and those forming a counterclockwise couple are negative. Thus, since $\tau_{\theta_{q1}}$ and $\tau_{\theta_{q2}}$ occur on mutually perpendicular planes, they must be of equal magnitude and opposite signs.

As in the case of principal stresses, the maximum and minimum shearing stresses determined by Equations (16) can be expressed purely in terms σ_x, σ_y, and τ_{xy}. By means of Equation (15) we can determine $\sin 2\theta_q$ and $\cos 2\theta_q$ as follows:

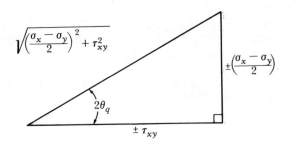

$$\sin 2\theta_q = \frac{\pm (\sigma_x - \sigma_y)/2}{\sqrt{[(\sigma_x - \sigma_y)/2]^2 + \tau_{xy}^2}}$$

$$\cos 2\theta_q = \frac{\pm \tau_{xy}}{\sqrt{[(\sigma_x - \sigma_y)/2]^2 + \tau_{xy}^2}}$$

After substituting $\sin 2\theta_q$ and $\cos 2\theta_q$ into Equation (13) and simplifying, we obtain

$$\tau_{\substack{max \\ min}} = \pm \sqrt{\left(\frac{\sigma_x - \sigma_y}{2}\right)^2 + \tau_{xy}^2} \tag{17}$$

Also note that one-half the difference between the principal stresses (Equation 11) is the maximum shearing stress:

$$\tau_{max} = \frac{(\sigma_{p1} - \sigma_{p2})}{2} \tag{18}$$

● **CAUTION** Usually we do not distinguish between maximum and minimum shearing stresses, since they are of equal magnitude and occur on mutually perpendicular planes. Henceforth, when we refer to maximum shearing stress, we shall mean values given by either Equation (16) or (17).

We have shown that there is no shearing stress on the principal planes. Let us now determine the normal stress (σ') on the planes of maximum shearing stress. If we substitute $\sin 2\theta_q$ and $\cos 2\theta_q$ into Equation (8) and simplify, we get

$$\sigma_{\theta_q} = \sigma' = \frac{\sigma_x + \sigma_y}{2} \tag{19a}$$

The invariant property of Equation (12) also permits us to write the normal stress σ' in terms of the principal stresses:

$$\sigma' = \frac{\sigma_{p1} + \sigma_{p2}}{2} . \tag{19b}$$

It is frequently desirable to show the principal stresses and the maximum shearing stress on rectangular elements (or on a single wedge-shaped element) properly oriented relative to the reference coordinate frame xy. Proper orientation of these elements will be illustrated in the examples in the Study Guide.

If you are somewhat confused by the large amount of material in this section, don't panic. It should become much clearer after we have examined some examples in Section SG2-3.

Review this section and then turn to the study program in Section SG2-3. ■ **STOP**

R2-4 Mohr's circle for stress

There is another technique for handling stress transformations and principal stress problems. Otto Mohr (1835–1918), a famous German mechanics professor, showed that the stress transformation equations are parametric equations of a circle. Mohr's circle can be used either as a purely graphical technique, or as an aid in the algebraic solution of problems.

▶ **NOTE** Mohr's circle can be a valuable aid for those students who develop a "feel" for the circle, and who prefer graphical aids in understanding mathematical relationships. Students who do not develop a feel for Mohr's circle may find it of little value, and may prefer to rely solely on the equations of Sections R2-2 and R2-3.

We shall proceed with the development of Mohr's circle. The stress transformation equations can be written in the form

$$\sigma_\theta - \left(\frac{\sigma_x + \sigma_y}{2}\right) = \left(\frac{\sigma_x - \sigma_y}{2}\right)\cos 2\theta - \tau_{xy}\sin 2\theta \qquad (20)$$

$$\tau_\theta = \left(\frac{\sigma_x - \sigma_y}{2}\right)\sin 2\theta + \tau_{xy}\cos 2\theta \qquad (21)$$

Squaring both sides of Equations (20) and (21) gives

$$\left[\sigma_\theta - \left(\frac{\sigma_x + \sigma_y}{2}\right)\right]^2 = \left(\frac{\sigma_x - \sigma_y}{2}\right)^2 \cos^2 2\theta \qquad (22)$$

$$- (\sigma_x - \sigma_y)\tau_{xy}\sin 2\theta \cos 2\theta + \tau_{xy}^2 \sin 2\theta$$

$$\tau_\theta^2 = \left(\frac{\sigma_x - \sigma_y}{2}\right)^2 \sin^2 2\theta \qquad (23)$$

$$+ (\sigma_x - \sigma_y)\tau_{xy}\sin 2\theta \cos 2\theta + \tau_{xy}^2 \cos^2 2\theta$$

After adding Equations (22) and (23), we obtain

$$\left[\sigma_\theta - \left(\frac{\sigma_x + \sigma_y}{2}\right)\right]^2 + \tau_\theta^2 = \left(\frac{\sigma_x - \sigma_y}{2}\right)^2 (\sin^2 2\theta + \cos^2 2\theta) \quad (24)$$

$$+ \tau_{xy}^2(\sin^2 2\theta + \cos^2 2\theta)$$

From trigonometric identities, we know that $\sin^2 2\theta + \cos^2 2\theta = 1$. Therefore Equation (24) simplifies to

$$\left[\sigma_\theta - \left(\frac{\sigma_x + \sigma_y}{2}\right)\right]^2 + (\tau_\theta - 0)^2 = \left(\frac{\sigma_x - \sigma_y}{2}\right)^2 + \tau_{xy}^2 \quad (25)$$

The general equation of a circle is of the form:

$$(x - a)^2 + (y - b)^2 = c^2 \quad (26)$$

where c is the radius and a and b are respectively the x and y coordinates of the center. Comparing Equations (25) and (26), we can see that, in a τ_θ-versus-σ_θ coordinate frame, Equation (25) is the equation of a circle centered at

$$(a, b) = \left(\frac{\sigma_x + \sigma_y}{2}, 0\right) \quad (27)$$

and with a radius of

$$r = \sqrt{\left(\frac{\sigma_x - \sigma_y}{2}\right)^2 + \tau_{xy}^2} \quad (28)$$

Figure 2-7 shows Mohr's circle for a state of stress in which σ_x, σ_y, and τ_{xy} are all positive and $\sigma_x > \sigma_y$. The coordinates of a point on the circle represent the normal and shearing stresses on a particular inclined plane.

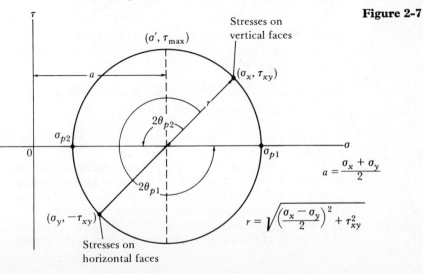

Figure 2-7

Two features of Mohr's circle frequently cause confusion: (1) What is the relationship between angles on Mohr's circle and angles on the element? (2) Why is σ_y plotted with $-\tau_{xy}$? We shall attempt to answer these questions in the next two paragraphs.

We showed earlier that the angle between principal stresses is 90°, but on Mohr's circle the stresses are diametrically opposite (180° apart). We also showed (in Section R2-3) that the planes of maximum shearing stress occur at 45° angles with respect to the principal planes, but on Mohr's circle they are 90° apart. Thus we must conclude that angles on a Mohr's circle are twice those in the physical plane. One further point: Our sign convention was chosen such that the direction of angular measurement on Mohr's circle would correspond to the direction of angular measurement on the element (i.e., clockwise angles on the element correspond to clockwise angles on Mohr's circle, etc.).

If we evaluate the transformation Equations (20) and (21) for $\theta = 0$, we get $(\sigma_\theta, \tau_\theta) = (\sigma_x, \tau_{xy})$. When we evaluate these equations for $\theta = 90°$, we get $(\sigma_\theta, \tau_\theta) = (\sigma_y, -\tau_{xy})$. Therefore, in our rotating coordinate frame defined by the orientation θ, the shearing stresses on mutually perpendicular planes must be of opposite signs. This conclusion is further borne out by the fact that our maximum shearing stress (Equation 17) occurs on mutually perpendicular planes, and has a positive value on one plane and a negative value on the other.

Let us now direct our attention once more to Mohr's circle (see Figure 2-7). The information obtained in Section R2-3 is available at once from the geometry of the circle.

a) $\sigma_{p1,\, p2} = a \pm r = \dfrac{\sigma_x + \sigma_y}{2} \pm \sqrt{\left(\dfrac{\sigma_x - \sigma_y}{2}\right)^2 + \tau_{xy}^2}$

b) $\tau_{\max} = r = \sqrt{\left(\dfrac{\sigma_x - \sigma_y}{2}\right)^2 + \tau_{xy}^2}$

c) The shearing stress on the principal planes is zero.
d) The normal stress on the planes of maximum shearing stress is $\sigma' = a = (\sigma_x + \sigma_y)/2$.

In summary, if we are given the state of stress defined by σ_x, σ_y, and τ_{xy}, we can utilize Mohr's circle as follows.

a) Sketch a coordinate frame with σ as the abscissa and τ as the ordinate.
b) Plot the points (σ_x, τ_{xy}) and $(\sigma_y, -\tau_{xy})$, which establish the diameter of the circle. (Remember the sign convention: Tensile stress is positive and compressive stress is negative; clockwise shearing couple τ_{xy} is positive, and counterclockwise is negative.)
c) Determine the desired quantities from the geometry of the circle. (The coordinates of a point on the circle represent the normal and

shearing stresses on a particular inclined plane. Keep in mind that the angles on the circle are measured in the same sense on the element, but have twice the magnitude on Mohr's circle.)

Turn to the study program of Section SG2-4. ■ **STOP**

R2-5   Absolute maximum shearing stress

In Section R2-3 we determined that, for a state of plane stress, the maximum shearing stress occurs on planes making 45° angles with the principal planes, and is given by $\tau_{\max} = (\sigma_{p1} - \sigma_{p2})/2$. The important exception occurs when the two principal stresses determined for the plane-stress state have the same sign (i.e., both positive or both negative). In that case, shearing stresses with magnitudes larger than $(\sigma_{p1} - \sigma_{p2})/2$ exist on planes which we have not yet considered.

The determination of "absolute maximum shearing stress" involves analysis of the triaxial state of stress. Although the study of the general triaxial state of stress and the accompanying transformations is beyond the scope of this course, some results from that theory are important in plane-stress problems, and will be given here without proof.

Advanced texts in mechanics of materials show that there always exist three principal planes (planes of maximum, minimum, and an intermediate normal stress) that are mutually perpendicular. Also the shearing stresses are always zero on these principal planes. In the case of plane stress, the third principal stress is always zero (that is, $\sigma_z = 0$ is the third principal stress).

Let us assume that the three principal stresses σ_{p1}, σ_{p2}, and σ_{p3} associated with a given point are known, and are shown on the three-dimensional elements in Figure 2-8. The absolute maximum shearing stress will act on two of the six 45° planes shown in the figure. Recall that the shearing stresses occur in pairs of equal magnitude but opposite directions on perpendicular planes.

Consider one of the 45° planes in Figure 2-8(a). We can determine the expression for the shearing stress $\tau'_{\max}$ acting on it by requiring that the sum of the forces in the tangential direction be equal to zero as follows:

$$\overset{+\searrow}{\sum} F_t = \tau'_{\max}\, dA - \sigma_{p1}\left(\frac{1}{\sqrt{2}}\, dA\right)\left(\frac{1}{\sqrt{2}}\right) + \sigma_{p2}\left(\frac{1}{\sqrt{2}}\, dA\right)\left(\frac{1}{\sqrt{2}}\right) = 0$$

and

$$\tau'_{\max} = \left|\frac{\sigma_{p1} - \sigma_{p2}}{2}\right| \tag{29a}$$

Similarly, the shear stresses on the other 45° planes shown in Figure 2-8(b) and (c) are

$$\tau''_{max} = \left| \frac{\sigma_{p3} - \sigma_{p2}}{2} \right| \tag{29b}$$

$$\tau'''_{max} = \left| \frac{\sigma_{p1} - \sigma_{p3}}{2} \right| \tag{29c}$$

The largest of the shearing stresses given by Equations (29a, b, c) is the absolute maximum shearing stress.

Figure 2-8

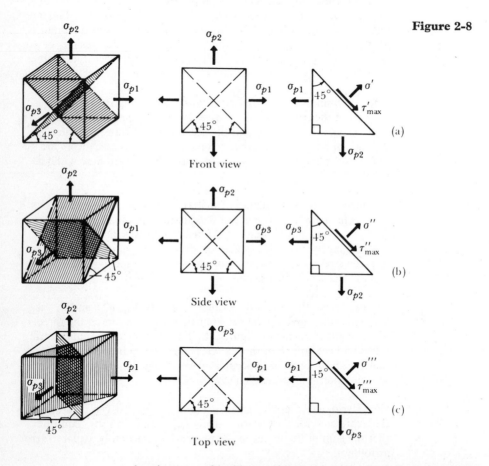

Another way of looking at this problem involves the use of Mohr's circle. A triaxial state of stress can be analyzed by the use of three Mohr's circles whose intercepts are the three principal stresses.

Consider the three possible plane-stress cases represented in Figure 2-9, in which each circle corresponds to the state of stress in one of the principal planes. The stresses σ_{p1} and σ_{p2} are the principal stresses given by plane stress theory, and $\sigma_{p3} = 0$ is the third principal stress. If σ_{p3} is the intermediate normal stress, we need not consider it. However, if σ_{p3} is either the maximum or the minimum normal stress, then it will combine with the other extreme value to give the absolute maximum stress. The absolute maximum shearing stress is always the radius of the largest of these three Mohr's circles.

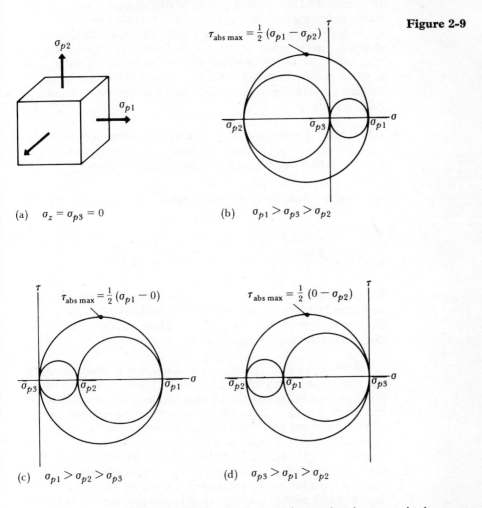

Figure 2-9

(a) $\sigma_z = \sigma_{p3} = 0$

(b) $\sigma_{p1} > \sigma_{p3} > \sigma_{p2}$

(c) $\sigma_{p1} > \sigma_{p2} > \sigma_{p3}$

(d) $\sigma_{p3} > \sigma_{p1} > \sigma_{p2}$

In summary, the absolute maximum shearing stress is always one-half the difference between the algebraically larger and smaller of the principal stresses (including the zero principal stress in the plane-stress problem).

Turn to the study program in Section SG2-5. ∎ **STOP**

R2-6   Strain transformation (plane stress)

In general, six independent strain values are required to completely define the state of strain at a point in a loaded body. These six strains can be the three normal strains and three shearing strains (ϵ_x, ϵ_y, ϵ_z, γ_{xy}, γ_{xz}, and γ_{yz}) corresponding to the triaxial stress state given in Figure 2-1 (Section R2-1).

In Section R2-1, we showed that three stresses σ_x, σ_y, and τ_{xy} completely define the state of plane stresses at a point, where $\sigma_z = \tau_{xz} = \tau_{yz} = 0$. If the shearing stresses τ_{xz} and τ_{yz} are zero, the shearing strains γ_{xz} and γ_{yz} must be zero (i.e., in linear elastic theory, $\tau_{xz} = G\gamma_{xz}$ and $\tau_{yz} = G\gamma_{yz}$). Therefore four strains (ϵ_x, ϵ_y, ϵ_z, and γ_{xy}) are required to completely define the state of strain at a point subjected to plane stress. However, ϵ_z can be written in terms of ϵ_x, ϵ_y, and Poisson's ratio μ.

● **CAUTION** The fact that $\sigma_z = 0$ in a plane-stress problem does *not* make ϵ_z zero. In Chapter 1 we derived the generalized Hooke's law, which shows that $\epsilon_z = (\sigma_z - \mu\sigma_x - \mu\sigma_y)/E$. Thus

$$\epsilon_z = \frac{-\mu}{E}(\sigma_x + \sigma_y) = \frac{-\mu}{1 - \mu}(\epsilon_x + \epsilon_y)$$

since

$$\epsilon_x + \epsilon_y = \frac{1 - \mu}{E}(\sigma_x + \sigma_y) \qquad \text{when} \qquad \sigma_z = 0$$

We can derive strain transformation laws analogous to the stress transformation laws of Section R2-2 from the geometry of the deformations of an element on a strained body. These transformation equations do not involve the normal strain ϵ_z.

We shall adopt a sign convention for strains that is compatible with the convention already adopted for stresses: Tensile strains (elongations) are positive, compressive strains (contractions) are negative, and shearing strains γ_{xy} resulting from positive shearing stresses τ_{xy} will be considered positive. A positive shearing stress τ_{xy}, as shown in Figure 2-10(a) will result in a deformed element, as illustrated by the positive γ_{xy} in Figure 2-10(b). Recall that shearing strain is the tangent of the angle of distortion (change from the original 90° angle), and it is approximated by the angle in radians for small angular deformations. A positive shearing strain results in a decrease in the lower right-hand angle of a rectangular element.

Figure 2-10

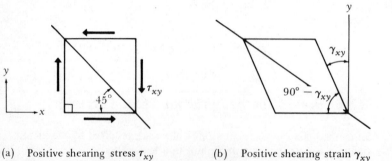

(a)   Positive shearing stress τ_{xy} (b)   Positive shearing strain γ_{xy}

To develop the strain transformation equations, we shall consider the geometric changes of a small rectangular element on a free face of a loaded body. Element $OABC$ in Figure 2-11(a) represents a rectangular element before the body is subjected to loading. After loading, the element will be deformed. Analysis of the effects of strain on the element will be somewhat easier to follow if we employ the principle of linear superposition and examine the effects of normal strains and shearing strain separately, then superimpose the results.

In Figure 2-11(b), we have an enlarged view of element $OABC$ deformed into $OA'B'C'$ by the normal strains ϵ_x and ϵ_y. These normal

Figure 2-11

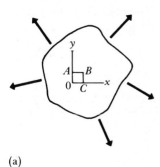

(a)

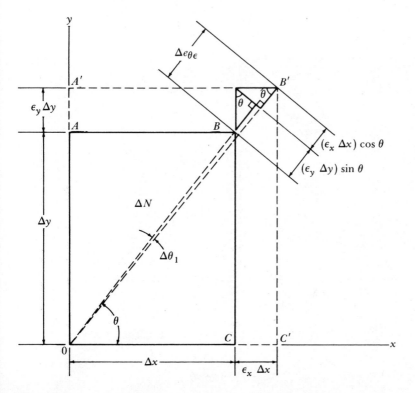

(b)

strains cause changes in the dimensions of the rectangular element, but not a change in shape (90° angles before loading remain 90° after loading). In the figure, all dimensional changes are greatly exaggerated to facilitate analysis. Keep in mind that all of the changes shown, both linear and angular, are very small (for example, $\epsilon_x \ll 1$, and $\Delta\theta_1 \ll \theta$ or $\theta - \Delta\theta_1 \cong \theta$). Deformations are shown to be to the right and upward, which causes no loss in generality, since we are interested in relative changes. The strains are assumed to be constant over the dimensions of the element. Therefore the dimensional changes are equal to the strains times their respective original lengths.

If we want to develop the expression for the normal strain at orientation θ, we need to know the elongation of line OB. The elongation of line OB resulting from the normal strains is

$$\Delta e_{\theta\epsilon} = (\epsilon_x \Delta x) \cos \theta + (\epsilon_y \Delta_y) \sin \theta \tag{30a}$$

In Figure 2-12, we have the deformed element $OA'B'C'$ distorted by shearing strain γ_{xy} into the new position $OA''B''C'$. The line OC' is held horizontal for reference purposes, which causes no loss in generality of the derivation.

Figure 2-12

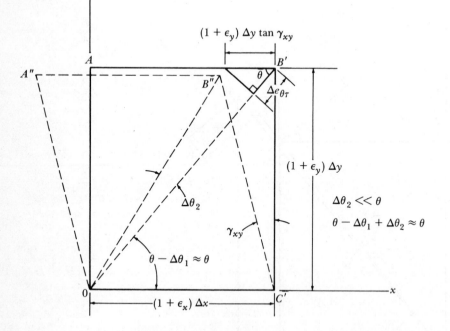

The change in the length of line OB' resulting from the positive shearing strain γ_{xy} is

$$\Delta e_{\theta\tau} = - [(1 + \epsilon_y) \Delta y \tan \gamma_{xy}] \cos \theta$$

where

$$\tan \gamma_{xy} \cong \gamma_{xy}, \qquad \epsilon_y \ll 1$$

Thus

$$\Delta e_{\theta_\tau} \cong -\gamma_{xy} \Delta y \cos \theta \tag{30b}$$

We can obtain the total elongation of line OB resulting from normal and shearing strains by adding their respective contributions, given by Equations (30a) and (30b).

$$\Delta e_\theta = \epsilon_x \Delta x \cos \theta + \epsilon_y \Delta y \sin \theta - \gamma_{xy} \Delta y \cos \theta \tag{31}$$

The normal strain ϵ_θ is, by definition, equal to the elongation of line OB divided by its original length ΔN.

$$\epsilon_\theta = \frac{\Delta e_\theta}{\Delta N} = \epsilon_x \left(\frac{\Delta x}{\Delta N}\right) \cos \theta + \epsilon_y \left(\frac{\Delta y}{\Delta N}\right) \sin \theta - \gamma_{xy} \left(\frac{\Delta y}{\Delta N}\right) \cos \theta \tag{32}$$

From Figure 2-11 we see that

$$\frac{\Delta x}{\Delta N} = \cos \theta \qquad \text{and} \qquad \frac{\Delta y}{\Delta N} = \sin \theta \tag{33}$$

Finally, substitution of Equations (33) into Equation (32) yields

$$\epsilon_\theta = \epsilon_x \cos^2 \theta + \epsilon_y \sin^2 \theta - \gamma_{xy} \sin \theta \cos \theta \tag{34}$$

Turn back to Section R2-2 and note the similarity between the normal-stress transformation equation (Equation 4) and the normal-strain transformation equation (Equation 34). ■ **STOP**

We could have obtained the strain transformation equation (Equation 34) by simply substituting the normal strain ϵ for the normal stress σ, and $\gamma/2$ for the shearing stress τ in Equation (4). This analogy holds for all stress- and strain-transformation equations. Therefore we can deduce the other strain-transformation equations from the corresponding stress-transformation equations.

We could, of course, consider the angular change in a set of rotated coordinates and obtain the transformation equation for shearing strain, or we could simply obtain it by substituting the appropriate symbols into the shearing-stress-transformation equation. We shall simply write the shearing-strain-transformation equation by analogy with the stress-transformation equation (Equation 5), substituting ϵ_x, ϵ_y, $\gamma_{xy}/2$, and $\gamma_\theta/2$ for σ_x, σ_y, τ_{xy}, and τ_θ, respectively:

$$\frac{\gamma_\theta}{2} = (\epsilon_x - \epsilon_y) \sin \theta \cos \theta + \frac{\gamma_{xy}}{2} (\cos^2 \theta - \sin^2 \theta) \tag{35}$$

The strain-transformation equations can also be written in the form

$$\epsilon_\theta = \frac{\epsilon_x + \epsilon_y}{2} + \left(\frac{\epsilon_x - \epsilon_y}{2}\right) \cos 2\theta - \frac{\gamma_{xy}}{2} \sin 2\theta \tag{36}$$

$$\gamma_\theta = (\epsilon_x - \epsilon_y) \sin 2\theta + \gamma_{xy} \cos 2\theta \tag{37}$$

Turn to the study program in Section SG2-6. ■ **STOP**

R2-7 Principal strains, Mohr's circle for strain

**Principal
strains**

When working with strains, as with stresses, we are usually interested in maximum and minimum values. We showed in Section R2-6 that the stress- and strain-transformation equations are of the same form. Thus equations for maximum and minimum values are also of the same form. The derivations of the equations for principal strains and maximum shearing strain are identical to those for stresses; only the names of the characters are changed. Therefore we shall not repeat those steps here.

The principal strains, maximum shearing strain, and their associated orientations are

$$\epsilon_{p1,\,p2} = \frac{\epsilon_x + \epsilon_y}{2} \pm \sqrt{\left(\frac{\epsilon_x - \epsilon_y}{2}\right)^2 + \left(\frac{\gamma_{xy}}{2}\right)^2} \tag{38}$$

$$\tan 2\theta_q = \frac{-\gamma_{xy}}{\epsilon_x - \epsilon_y} \tag{39}$$

and

$$\frac{\gamma_{\substack{max \\ min}}}{2} = \pm \sqrt{\left(\frac{\epsilon_x - \epsilon_y}{2}\right)^2 + \left(\frac{\gamma_{xy}}{2}\right)^2} \tag{40}$$

$$\tan 2\theta_q = \frac{\epsilon_x - \epsilon_y}{\gamma_{xy}} \tag{41}$$

▶ **NOTE** All our previous conclusions about stresses apply to strains (e.g., the principal-strain orientations are 90° apart, the maximum shearing strains occur on planes that are 45° from the principal planes, the shearing strain associated with the principal-strain orientations is zero, and so forth.)

**Mohr's circle
for strain**

Mohr's circle for strains is also analogous to Mohr's circle for stresses. Since the development is identical to that for stresses in Section R2-4, we shall not repeat it here.

● **CAUTION** Remember that we can obtain the strain transformation equations from the stress transformation equations by replacing σ with

ϵ and τ with $\gamma/2$. This means that the ordinate on Mohr's circle for strain is $\gamma/2$ and the abscissa is ϵ.

Figure 2-13 illustrates Mohr's circle for strain.

Figure 2-13

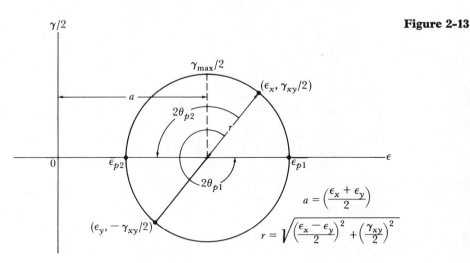

$$a = \left(\frac{\epsilon_x + \epsilon_y}{2}\right)$$

$$r = \sqrt{\left(\frac{\epsilon_x - \epsilon_y}{2}\right)^2 + \left(\frac{\gamma_{xy}}{2}\right)^2}$$

Turn to the study program in Section SG2-7. ■ **STOP**

R2-8   Strain rosettes, strain measurement

In previous sections of this chapter we assumed that somehow we had obtained the state of stress (σ_x, σ_y, τ_{xy}) or strain (ϵ_x, ϵ_y, γ_{xy}) at a point on a free surface of a loaded body. Stress and strain at a point, however, are analytical abstractions. In actuality we never measure either stress or strain at a point. Only in the simplest loading situations can we even calculate the stresses after measuring the loads and the geometry of the body.

Practical problems usually require that we measure strains over finite length in a physical test and then relate them to stresses. Although we cannot measure strain at a point, we can measure strain (or deformation) over a relatively short finite length. We then assume that the average strain over that length is representative of the strain at a point on or near the line of measurement.

There are many devices for measuring normal strains or deformations. The electrical-resistance strain gage is the most widely used device for normal strain measurement. Its operation is based on the principle that a change in the electrical resistance of a wire is directly proportional to the axial deformation of the wire. Commercially available resistance strain gages are usually made of parallel elements of wire or foil attached to a carrier (e.g., paper or bakelite). These gages are sensitive primarily to deformations parallel to the elements (i.e., they respond to normal strain or deformation in one direction). When the gage is properly cemented to a specimen, it is insulated from the specimen by the carrier and the cement, and will undergo the same deformation or strain as the surface of the specimen to which it is attached. Through properly calibrated instruments (strain indicators), which respond to changes in resistance, we can determine the accompanying normal strains.

There are also devices for measuring shearing strains, but they are usable only in restricted, very simple loading situations.

To determine the state of strain (ϵ_x, ϵ_y, γ_{xy}) at a point, we usually measure normal strains at three arbitrary orientations (a), (b), and (c), as shown in Figure 2-14. If we substitute the known measured normal strains ϵ_a, ϵ_b, and ϵ_c and their respective orientations θ_a, θ_b, and θ_c into the strain-transformation equation (Equation 34), we obtain Equations (42).

$$\epsilon_a = \epsilon_x \cos^2 \theta_a + \epsilon_y \sin^2 \theta_a - \gamma_{xy} \sin \theta_a \cos \theta_a$$

$$\epsilon_b = \epsilon_x \cos^2 \theta_b + \epsilon_y \sin^2 \theta_b - \gamma_{xy} \sin \theta_b \cos \theta_b \qquad (42)$$

$$\epsilon_c = \epsilon_x \cos^2 \theta_c + \epsilon_y \sin^2 \theta_c - \gamma_{xy} \sin \theta_c \cos \theta_c$$

Figure 2-14

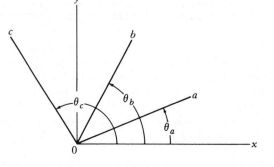

Equations (42) can be solved simultaneously for the three unknown quantities ϵ_x, ϵ_y, and γ_{xy}. In this way, we can determine the state of strain at a point by measuring only normal strains, avoiding the necessity of measuring shearing strains. Once we know ϵ_x, ϵ_y, and γ_{xy} at a particular point, we can determine the state of stress (σ_x, σ_y, τ_{xy}), the principal strains and maximum shearing strain, the principal stresses, and maximum shearing stress, and the angles associated with these stresses and strains at that point.

As an example, let us examine a common gage configuration. Assume that the three electrical-resistance strain gages are oriented at angles of 0°, 45°, and 90° with respect to the x axis, as illustrated in Figure 2-15. Equations (42) simplify as follows:

$$\epsilon_a = \epsilon_0 = \epsilon_x \overset{1}{\cancel{\cos^2 0°}} + \epsilon_y \overset{0}{\cancel{\sin^2 0°}} - \gamma_{xy} \overset{0}{\cancel{\sin 0°}} \cos 0°$$

Therefore $\epsilon_a = \epsilon_0 = \epsilon_x$

$$\epsilon_c = \epsilon_{90} = \epsilon_x \overset{0}{\cancel{\cos^2 90°}} + \epsilon_y \overset{1}{\cancel{\sin^2 90°}} - \gamma_{xy} \sin 90° \overset{0}{\cancel{\cos 90°}}$$

Therefore $\epsilon_c = \epsilon_{90} = \epsilon_y$ and

$$\epsilon_b = \epsilon_{45} = \epsilon_x \cos^2 45° + \epsilon_y \sin^2 45° - \gamma_{xy} \sin 45° \cos 45°$$

$$= \epsilon_x \left(\frac{1}{\sqrt{2}}\right)^2 + \epsilon_y \left(\frac{1}{\sqrt{2}}\right)^2 - \gamma_{xy} \left(\frac{1}{\sqrt{2}}\right) \left(\frac{1}{\sqrt{2}}\right)$$

$$\gamma_{xy} = \epsilon_x + \epsilon_y - 2\epsilon_b \quad \text{or} \quad \gamma_{xy} = \epsilon_0 + \epsilon_{90} - 2\epsilon_{45} = \epsilon_a + \epsilon_c - 2\epsilon_b$$

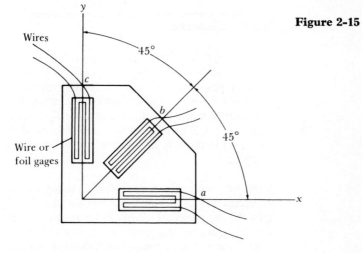

Figure 2-15

We can reduce the amount of work involved in solving Equations (42) by choosing our reference axis x to coincide with the orientation of the first gage. Then there are at most two coupled simultaneous equations to be solved.

Another commonly used gage configuration is the delta, which has gages oriented 60° apart (that is, 0°, 60°, 120°), as illustrated in Figure 2-16. Devices for measuring normal strains at two or more orientations, such as those shown in Figures 2-15 and 2-16, are called rosette strain gages.

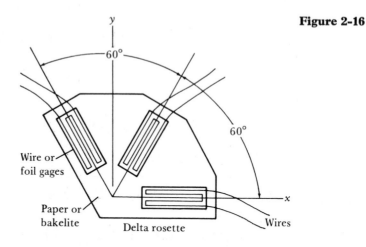

Figure 2-16

Delta rosette

Turn to the study program in Section SG2-8. ■ **STOP**

R2-9   Closure

In this chapter, we first developed stress-transformation equations and methods to determine principal stresses, maximum shearing stress, and their respective orientations. Next, we studied strain transformations, principal strains, maximum shearing strain, and their associated orientations. Finally, we examined rosette equations and strain measurement. The material was thus developed in the reverse order compared to the way in which it is usually applied in a practical problem.

From measurements of three normal strains, we should now be able to determine: (a) the state of strain associated with the xy coordinate axes, (b) the principal strains and maximum shearing strain, (c) the state of stress, (d) the principal stresses and maximum shearing stress, and (e) the angles associated with these stresses and strains.

The stress and strain transformation equations combined with the generalized Hooke's law (Section R1-3) are the only tools we need. The accompanying flow sheet shows two ways to reduce data on strain rosettes to principal stresses. Of course, all the operations can be performed in the reverse of the order indicated. However, experimental problems usually start with strain rosette data and require the calculation of principal stresses and strains.

After studying the following flow sheet, turn to Section SG2-9.
■ **STOP**

Principal stresses from strain rosette data
(Plane stress: $\sigma_z = \tau_{xz} = \tau_{yz} = 0$)

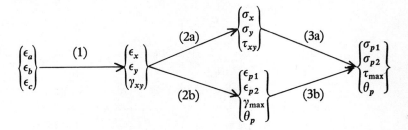

Equations for step 1

$$\epsilon_a = \epsilon_x \cos^2 \theta_a + \epsilon_y \sin^2 \theta_a - \gamma_{xy} \sin \theta_a \cos \theta_a$$

$$\epsilon_b = \epsilon_x \cos^2 \theta_b + \epsilon_y \sin^2 \theta_b - \gamma_{xy} \sin \theta_b \cos \theta_b$$

$$\epsilon_c = \epsilon_x \cos^2 \theta_c + \epsilon_y \sin^2 \theta_c - \gamma_{xy} \sin \theta_c \cos \theta_c$$

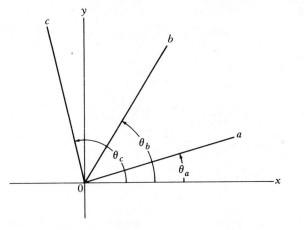

Equations for step 2a

$$\sigma_x = \frac{E}{(1 - \mu^2)} (\epsilon_x + \mu\epsilon_y)$$

$$\sigma_y = \frac{E}{(1 - \mu^2)} (\epsilon_y + \mu\epsilon_x)$$

$$\tau_{xy} = G\gamma_{xy}$$

Equations for step 2b

$$\epsilon_{p1,p2} = \frac{\epsilon_x + \epsilon_y}{2} \pm \sqrt{\left(\frac{\epsilon_x - \epsilon_y}{2}\right)^2 + \left(\frac{\gamma_{xy}}{2}\right)^2}$$

$$\frac{\gamma_{max}}{2} = \sqrt{\left(\frac{\epsilon_x - \epsilon_y}{2}\right)^2 + \left(\frac{\gamma_{xy}}{2}\right)^2} = \frac{\epsilon_{p1} - \epsilon_{p2}}{2}$$

$$\tan 2\theta_p = \frac{-\gamma_{xy}}{\epsilon_x - \epsilon_y}$$

$$\epsilon_\theta = \frac{\epsilon_x + \epsilon_y}{2} + \left(\frac{\epsilon_x - \epsilon_y}{2}\right) \cos 2\theta - \frac{\gamma_{xy}}{2} \sin 2\theta$$

Equations for step 3a (same as 2b with $\epsilon \to \sigma$ and $\gamma \to 2\tau$)

$$\sigma_{p1,p2} = \frac{\sigma_x + \sigma_y}{2} \pm \sqrt{\left(\frac{\sigma_x - \sigma_y}{2}\right)^2 + \tau_{xy}^2}$$

$$\tau_{max} = \sqrt{\left(\frac{\sigma_x - \sigma_y}{2}\right)^2 + \tau_{xy}^2} = \frac{\sigma_{p1} - \sigma_{p2}}{2}$$

$$\tan 2\theta_p = \frac{-2\tau_{xy}}{\sigma_x - \sigma_y}$$

$$\sigma_\theta = \frac{\sigma_x + \sigma_y}{2} + \left(\frac{\sigma_x - \sigma_y}{2}\right) \cos 2\theta - \tau_{xy} \sin 2\theta$$

Equations for step 3b (same as 2a except for subscripts)

$$\sigma_{p1} = \frac{E}{1 - \mu^2} (\epsilon_{p1} + \mu\epsilon_{p2})$$

$$\sigma_{p2} = \frac{E}{1 - \mu^2} (\epsilon_{p2} + \mu\epsilon_{p1})$$

$$\tau_{max} = G\gamma_{max}$$

$$\theta_p(\text{strain}) = \theta_p(\text{stress})$$

Problems

2-1.1 A 4 in. × 8 in. wooden block 8 in. high carries a uniformly distributed load of 16,000 pounds. Use the wedge method of analysis to determine the normal and shearing stress on the plane of the grain.

2-1.2 through 2-1.4 Use the wedge method to determine the normal and shearing stresses on the inclined planes for the plane-stress states shown in the figures. (Figure 2-1.4 is on page 54.)

2-1.5 The resultant of a uniformly distributed stress on the inclined plane of the block shown is 1300 pounds. Determine the normal and shearing stresses on the inclined plane; also determine σ_x and σ_y. (*Note:* $\tau_{xy} = 0$, and R, σ_x, and σ_y all lie in a common plane.) (See Figure 2-1.5 on page 54.)

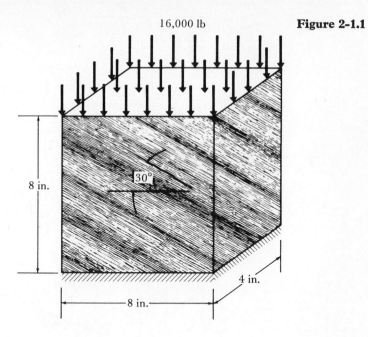

16,000 lb

8 in.

30°

8 in.

4 in.

Figure 2-1.1

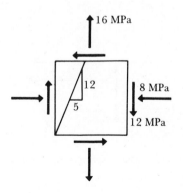

16 MPa

12

5

8 MPa

12 MPa

Figure 2-1.2

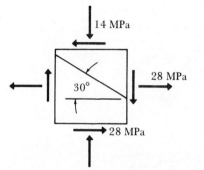

14 MPa

30°

28 MPa

28 MPa

Figure 2-1.3

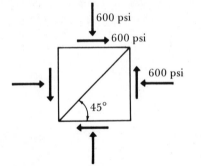

Figure 2-1.4

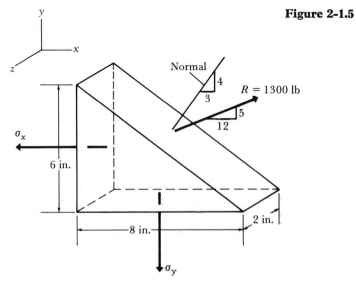

Figure 2-1.5

2-1.6 Determine the unknown stresses σ_x, τ_{xy}, and τ_{yx} on the wedge element shown.

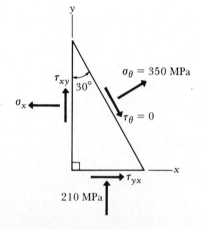

Figure 2-1.6

2-1.7 Write equations for σ_θ and τ_θ in terms of the angle θ and the axial stress σ_x. Obtain the maximum and minimum values of σ_θ and τ_θ by setting $d\sigma_\theta/d\theta = 0$ and $d\tau_\theta/d\theta = 0$. Use the accompanying figure.

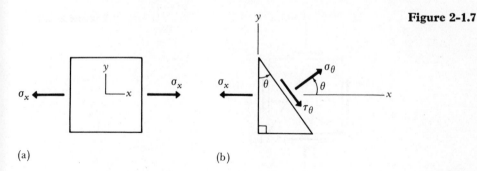

Figure 2-1.7

(a)   (b)

2-1.8 Specifications for the timber block shown require that the stresses not exceed 1.2-MPa shearing stress parallel to the grain, and 4.8-MPa compressive stress normal to the grain. Determine the maximum compressive load P that the block may carry. Assume that P is uniformly distributed over the cross section.

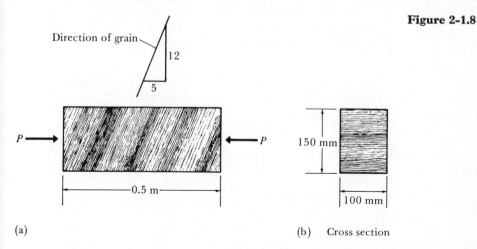

Figure 2-1.8

(a)   (b)   Cross section

2-2.1 through 2-2.4 Rework Problems 2-1.1 through 2-1.4, using the transformation equations.

2-2.5 Use the transformation equations to determine the shearing and normal stresses on a plane oriented 22.5° counterclockwise from the horizontal when the state of plane stress is given by:

$$\sigma_x = +20 \text{ MPa}, \qquad \sigma_y = -10 \text{ MPa}, \qquad \tau_{xy} = -15 \text{ MPa}$$

2-2.6 Use the transformation equations to determine the normal and shearing stresses on planes oriented 30° counterclockwise from the horizontal and vertical planes. Show your results on a rectangular element oriented 30° counterclockwise from the horizontal. Refer to the accompanying figure.

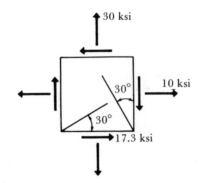

Figure 2-2.6

2-2.7 Replace θ by $\theta + 90°$ in the normal-stress-transformation equation. Then show that: $\sigma_\theta + \sigma_{\theta+90°} = \sigma_x + \sigma_y$. Is this relationship satisfied in Problem 2-2.6?

2-3.1 through 2-3.4 For the states of plane stress shown, determine the principal stresses, the maximum shearing stress, and show them on properly oriented elements (or on a single wedge-shaped element). Also show the normal stress on the planes of maximum shearing stress.

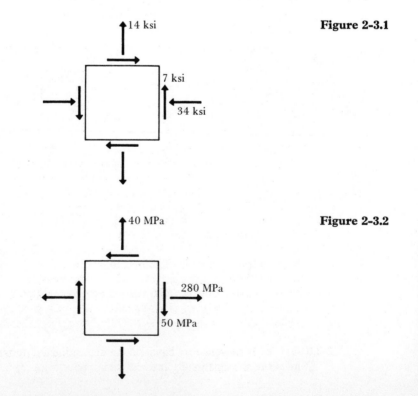

Figure 2-3.1

Figure 2-3.2

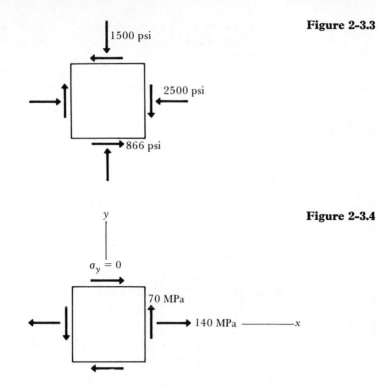

Figure 2-3.3

Figure 2-3.4

In each of the following plane-stress problems, 2-3.5 through 2-3.8, determine the principal stresses, the maximum shearing stress, and show these stresses on properly oriented elements. Be sure to show the normal stress on the planes of maximum shearing stress.

2-3.5 $\sigma_x = 0$, $\qquad$ $\sigma_y = 6000$ psi, $\qquad$ $\tau_{xy} = -4000$ psi

2-3.6 $\sigma_x = -10.0$ MPa, $\qquad$ $\sigma_y = -20.0$ MPa, $\qquad$ $\tau_{xy} = -8.66$ MPa

2-3.7 $\sigma_x = 8$ ksi, $\qquad$ $\sigma_y = -6$ ksi, $\qquad$ $\tau_{xy} = +4$ ksi

2-3.8 $\sigma_x = -200$ MPa, $\qquad$ $\sigma_y = 280$ MPa, $\qquad$ $\tau_{xy} = +70$ MPa

2-4.1 through 2-4.4 Rework Problems 2-3.1 through 2-3.4, using Mohr's circle as an aid in the algebraic solutions.

2-4.5 Sketch Mohr's circle for each of the three stress states shown. Determine the principal stress, the maximum shearing stress, and show them on properly oriented elements.

2-4.6 Sketch Mohr's circle for the state of stress shown, and then use it to determine the state of stress on a rectangular element oriented at 45° to the one shown.

2-4.7 and 2-4.8 Rework Problems 2-1.2 and 2-1.3, using Mohr's circle as an aid in the algebraic solutions.

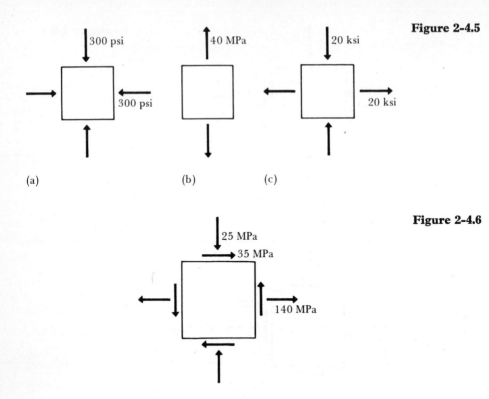

Figure 2-4.5

(a)   (b)   (c)

Figure 2-4.6

2-5.1 through 2-5.4 Determine the absolute maximum shearing stress for the plane-stress states given in Problems 2-3.1 through 2-3.4.

2-5.5 Determine the absolute maximum shearing stress for the plane-stress state given in Problem 2-4.5(a). If σ_z were 300 psi compressive in this problem, what would the absolute maximum shearing stress be?

2-5.6 through 2-5.8 Determine the absolute maximum shearing stress for the plane-stress states given in Problems 2-4.6 through 2-4.8.

2-6.1 Replace θ by $\theta + 90°$ in the normal-strain transformation equation; then show that: $\epsilon_x + \epsilon_y = \epsilon_\theta + \epsilon_{\theta+90°}$

2-6.2 A strain gage is mounted as shown at point A on a uniaxial test specimen. The specimen is loaded and the gage reads 0.001 m/m strain. Given that the modulus of elasticity is 140 GPa and Poisson's ratio is $\frac{1}{3}$, what is the axial strain (i.e., the strain in the direction of the load P)? If the rectangular cross section of the specimen is 2 cm × 4 cm, what is the axial load P?

2-6.3 The normal strains at a point are

$$\epsilon_a = -5 \times 10^{-3} \text{ m/m}, \qquad \epsilon_b = 15 \times 10^{-3} \text{ m/m},$$
$$\epsilon_c = 15 \times 10^{-3} \text{ m/m}$$

Determine ϵ_d and ϵ_e. Refer to the accompanying figure on page 59.

Figure 2-6.2

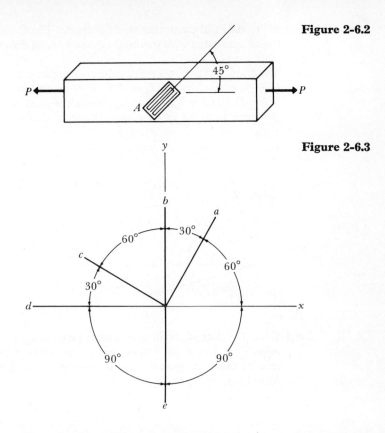

Figure 2-6.3

2-6.4 Determine ϵ_u for the uniaxial loading state shown. Assume that the load is uniformly distributed over the cross section. (*Hint:* $\gamma_{xy} = \gamma_{xz} = 0$) $E = 10 \times 10^6$ psi, $\mu = \frac{1}{3}$.

Figure 2-6.4

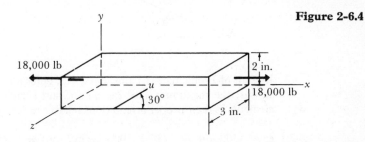

2-6.5 For the loading state and specimen shown in Problem 2-6.4, determine the angle θ (measured from the direction of loading) for which the normal strain is zero.

2-6.6 The circumferential strain on the surface of a cylindrical pressure vessel is 500×10^{-6} m/m. Utilizing the biaxial Hooke's law and the thin-walled pressure vessel theory from Chapter 1, show that the axial normal strain should be 100×10^{-6} m/m if the material is aluminum (Poisson's ratio = $\frac{1}{3}$). Also determine the normal strain at

45° to the axial-circumferential directions. The shearing strain (and stress) associated with the axial-circumferential directions is zero.

2-6.7 The state of strain at a point subjected to plane stress is given by $\epsilon_x = 5 \times 10^{-3}$ in./in., $\epsilon_y = 3 \times 10^{-3}$ in./in., and $\gamma_{xy} = 4 \times 10^{-3}$ in./in. Determine ϵ_u, ϵ_v, and γ_{uv}, where the uv coordinate frame is 15° clockwise from the xy coordinate frame, as shown.

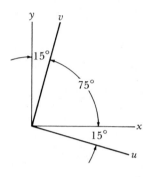

Figure 2-6.7

2-6.8 When the state of strain at a point is given as $\epsilon_x = -800 \times 10^{-6}$ m/m, $\epsilon_y = 300 \times 10^{-6}$ m/m, and $\gamma_{xy} = -800 \times 10^{-6}$ m/m, determine the orientation θ of the uv coordinate frame for which $\gamma_{uv} = 0$. Also determine ϵ_u and ϵ_v.

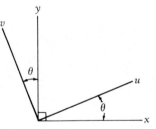

Figure 2-6.8

In each of Problems 2-7.1 through 2-7.6, determine the principal strains, the maximum shearing strain, and sketch the deformed elements in their proper orientations relative to the xy coordinate frame.

2-7.1 $\epsilon_x = 1500$ μin./in., $\epsilon_y = 500$ μin./in., $\gamma_{xy} = -2400$ μin./in.

2-7.2 $\epsilon_x = -600 \times 10^{-6}$ m/m, $\epsilon_y = 600 \times 10^{-6}$ m/m, $\gamma_{xy} = -1600 \times 10^{-6}$ m/m

2-7.3 $\epsilon_x = 1000$ μin./in., $\epsilon_y = -600$ μin./in., $\gamma_{xy} = 1200$ μin./in.

2-7.4 $\epsilon_x = -40 \times 10^{-3}$ m/m, $\epsilon_y = 60 \times 10^{-3}$ m/m, $\gamma_{xy} = 24 \times 10^{-3}$ m/m

2-7.5 $\epsilon_x = -30 \times 10^{-3}$ in./in., $\epsilon_y = -10 \times 10^{-3}$ in./in., $\gamma_{xy} = -20 \times 10^{-3}$ in./in.

2-7.6 $\epsilon_x = -140 \times 10^{-6}$ m/m, $\epsilon_y = 0$, $\gamma_{xy} = -480 \times 10^{-6}$ m/m

2-7.7 The principal strains at a point are:

$$\epsilon_{max} = \epsilon_u = 600 \times 10^{-6} \text{ in./in.}, \qquad \epsilon_{min} = \epsilon_v = -1400 \times 10^{-6} \text{ in./in.}$$

Determine ϵ_x, ϵ_y, and γ_{xy}, given that the orientations u and v are 15° clockwise from x and y, respectively. See the accompanying figure.

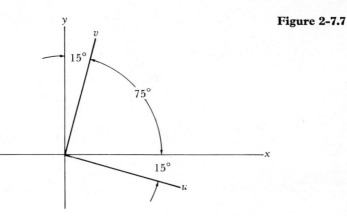

Figure 2-7.7

2-8.1 Given:

$$\epsilon_a = 2500 \times 10^{-6} \text{ in./in.}, \qquad \epsilon_b = 3600 \times 10^{-6} \text{ in./in.},$$
$$\epsilon_c = -500 \times 10^{-6} \text{ in./in.}$$

Determine ϵ_x, ϵ_y, γ_{xy}. See the accompanying figure.

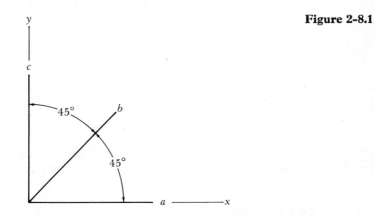

Figure 2-8.1

2-8.2 Given:

$$\epsilon_a = 20 \times 10^{-4} \text{ in./in.}, \qquad \epsilon_b = -12 \times 10^{-4} \text{ in./in.},$$

$$\epsilon_c = 2 \times 10^{-4} \text{ in./in.}$$

Determine ϵ_x, ϵ_y, and γ_{xy}, as well as the principal strains and the maximum shearing strain. Give the orientation of the principal strains relative to orientation a. Refer to the figure.

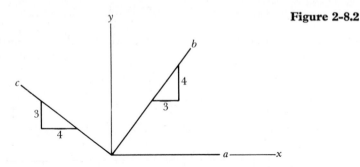

Figure 2-8.2

2-8.3 Given the normal strain data of Problem 2-6.3, determine the principal strains and their orientations.

2-8.4 Given:

$$\epsilon_{0°} = -2 \times 10^{-3} \text{ in./in.}, \qquad \epsilon_{30°} = 0, \qquad \epsilon_{90°} = 6 \times 10^{-3} \text{ in./in.}$$

Determine the principal strains and their orientations. Note that this is equivalent to a uniaxial loading state when $\mu = \frac{1}{3}$. If $E = 30 \times 10^6$ psi, what are the principal stresses? See the accompanying figure.

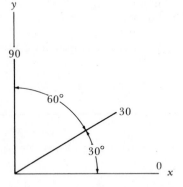

Figure 2-8.4

The data for Problems 2-8.5 through 2-8.7 refer to the orientations shown in the figure. Determine the magnitude and direction of the principal strains in each case.

2-8.5 $\epsilon_a = 1000 \ \mu\text{m/m}$, $\epsilon_c = 400 \ \mu\text{m/m}$, $\epsilon_e = 400 \ \mu\text{m/m}$

2-8.6 $\epsilon_a = 110 \ \mu\text{m/m}$, $\epsilon_b = 285 \ \mu\text{m/m}$, $\epsilon_d = -140 \ \mu\text{m/m}$

2-8.7 $\epsilon_a = -250 \ \mu\text{m/m}$, $\epsilon_c = 300 \ \mu\text{m/m}$, $\epsilon_e = -200 \ \mu\text{m/m}$

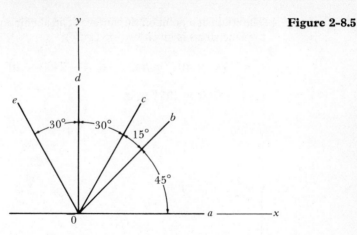

Figure 2-8.5

2-9.1 Given the strain in Problem 2-8.1, determine the stresses σ_x, σ_y, and τ_{xy}, and from them determine the principal stresses. Show the principal stresses on a properly oriented element. ($E = 30 \times 10^6$ psi, $G = 12 \times 10^6$ psi)

2-9.2 Rework Problem 2-9.1 by solving for the principal strains and the maximum shearing strain. Next determine the principal stresses and the maximum shearing stress. Show these stresses on properly oriented elements.

2-9.3 The state of strain on the surface of a loaded steel specimen is (see figure):

$$\epsilon_a = -130 \times 10^{-5} \text{ m/m}, \qquad \epsilon_b = 45 \times 10^{-5} \text{ m/m},$$

$$\epsilon_c = 70 \times 10^{-5} \text{ m/m}$$

Figure 2-9.3

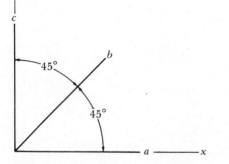

Determine σ_x, σ_y, and τ_{xy}, and then determine the principal stresses. Show the principal stresses on a properly oriented element. The elastic constants are: $E = 200$ GPa, $\mu = \frac{1}{4}$, $G = 80$ GPa.

2-9.4 The strain at a point on the surface of an aluminum specimen subjected to plane stress is given by (see figure):

$$\epsilon_a = 4400 \times 10^{-6} \text{ m/m}, \qquad \epsilon_b = -2400 \times 10^{-6} \text{ m/m},$$

$$\epsilon_c = -5200 \times 10^{-6} \text{ m/m}$$

Figure 2-9.4

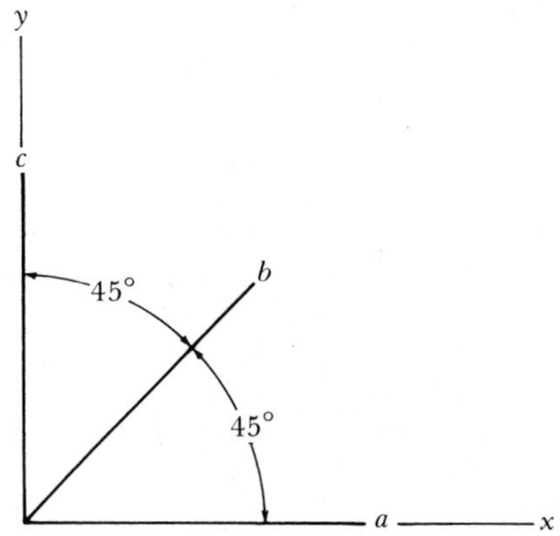

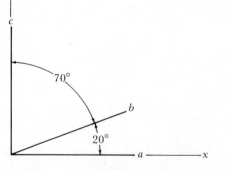

Determine the principal strains and principal stresses. Show the principal stresses on a properly oriented element. The elastic constants are: $E = 70$ GPa, $\mu = \frac{1}{3}$.

2-9.5 The normal strains at a point are (see figure):

$$\epsilon_a = 4 \times 10^{-4} \text{ in./in.}, \qquad \epsilon_b = 4 \times 10^{-4} \text{ in./in.},$$

$$\epsilon_c = 6 \times 10^{-4} \text{ in./in.}$$

Figure 2-9.5

Determine the principal strains and principal stresses. Show the principal stresses on a properly oriented element. The elastic constants are: $E = 30 \times 10^6$ psi, $\mu = \frac{1}{4}$.

2-9.6 A delta rosette at a point on the surface of a loaded body indicates the following strains (see figure):

$$\epsilon_a = 140 \times 10^{-6} \text{ m/m}, \qquad \epsilon_b = 212 \times 10^{-6} \text{ m/m},$$

$$\epsilon_c = -307 \times 10^{-6} \text{ m/m}$$

Figure 2-9.6

Determine σ_x, σ_y, τ_{xy} and then calculate the principal stresses and show them on a properly oriented element. ($E = 200$ GPa, $\mu = \frac{1}{4}$, $G = 80$ GPa.)

2-9.7 The strains on the surface of an aluminum member were experimentally found to be (see figure):

$$\epsilon_a = -3000 \times 10^{-6} \text{ m/m}, \qquad \epsilon_b = -3000 \times 10^{-6} \text{ m/m},$$

$$\epsilon_c = 2200 \times 10^{-6} \text{ m/m}$$

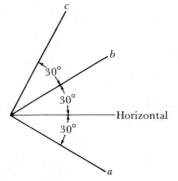

Figure 2-9.7

Determine the principal strains and principal stresses. Show the principal stresses on a properly oriented element. ($E = 70$ GPa, $\mu = \frac{1}{3}$.)

2-9.8 The strains on the outside surface of a thin-walled cylindrical pressure vessel are (see figure):

$$\epsilon_{\text{axial}} = 400 \times 10^{-6} \text{ in./in.}, \qquad \epsilon_{45°} = 900 \times 10^{-6} \text{ in./in.},$$

$$\epsilon_{\text{hoop}} = 1400 \times 10^{-6} \text{ in./in.}$$

where $E = 30 \times 10^6$ psi, $\mu = \frac{1}{4}$, $t = \frac{1}{10}$ in. (wall thickness), $r_i =$ 15. in. (inside radius). Determine the internal pressure in the vessel.

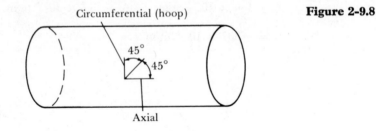

Circumferential (hoop)

45°

45°

Axial

Figure 2-9.8

Chapter three
Torsion

R3-1 Torsional shear strain and stress

In the study of mechanics of materials, we depend on assumptions and geometric observations in our analysis of relatively simple loading situations. In this section, we shall analyze the strain and stress distributions on circular torsion members and determine how they are related to torque. The analytical solution to this problem depends on the following assumptions.

a) Plane cross sections before twisting remain plane after twisting, and the diameter of the cross section remains a straight line after twisting (this is true only for circular sections).
b) All longitudinal elements must have the same length (which is true only for straight shafts with constant diameters).
c) The modulus of rigidity G must have the same value at all points and in all directions (which implies that the material must be homogeneous, isotropic, and linearly elastic).

Assumption (a) can be experimentally verified, and is found to be valid not only in the elastic range, but well into the plastic range of deformations. These geometric observations permit us to determine how the shearing strain is distributed on the cross section.

Figure 3-1

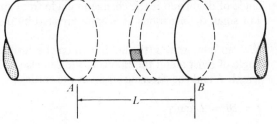

Consider a member with a circular cross section (Figure 3-1) on which we have drawn two reference circles A and B separated by a distance L. We shall also draw a horizontal line and a small rectangular element on the surface. After the application of torque, section B will rotate relative to section A, and the horizontal line will form a cylindrical helix around the specimen. The rectangular element will also be distorted. Figure 3-2 shows the initial and deformed elements. For reference purposes, we can assume that section A does not rotate—an assumption that causes no loss in generality, since we are concerned only with relative rotations.

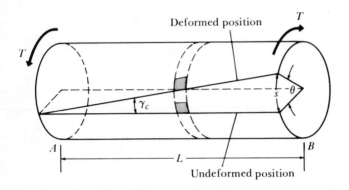

Figure 3-2

(a)

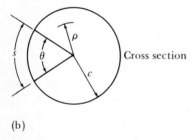

Cross section

(b)

If a surface layer could be peeled off the segment and flattened, it would look like the element in Figure 3-3. The angular distortion (shear strain) associated with the rectangular element is γ_c. Recall that the shearing strain is the tangent of the angle of distortion (the angle of distortion is the change in angle from an original 90° angle). For small deformations, it is approximated by the angle in radians.

We can now write s (the arc length on the surface) in terms of the angle of twist θ and the radius c. We can also write s in terms of the length L and the shearing strain at the surface, γ_c:

$$s = c\theta = L \tan \gamma_c \qquad (1)$$

In most engineering applications, we assume small deformations, and thus can make the approximation $\tan \gamma_c \cong \gamma_c$. Therefore

$$\gamma_c = \frac{c\theta}{L} \qquad (2)$$

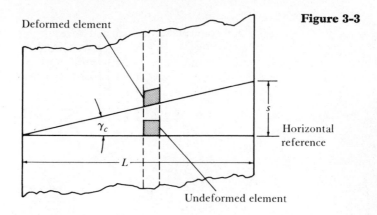

Figure 3-3

If we examine a similar section at radial distance ρ from the center of the specimen, we can write an expression for the arc length s_ρ at a distance ρ from the center as follows:

$$s_\rho = \rho\theta = L \tan \gamma \cong L\gamma$$

where γ is the shearing strain at a distance ρ from the center. Solving for γ, we obtain

$$\gamma = \frac{\rho\theta}{L} \tag{3}$$

We see from Equations (2) and (3) that the shearing strain varies linearly from the center of the specimen and has its maximum value at the surface (where $\rho = c$).

▶ **NOTE** The deformed elements of Figures 3-2 and 3-3, as well as Equations (1), (2), and (3), depend for their validity on assumptions (a) and (b).

If we assume linear elastic theory (assumption c), then we find that the shear stress is directly proportional to the shear strain, and we can write

$$\tau = G\gamma = \frac{G\rho\theta}{L} = k\rho \tag{4}$$

where

$$k = \frac{G\theta}{L} = \frac{\tau}{\rho}$$

From Equation (4), we see that the shearing stress is directly proportional to the radial distance. Thus we can now determine the relationship between torque and shear stress. We can obtain the torque on the section by summing the moments about the center of the cross section, as illustrated in Figure 3-4 and Equation (5).

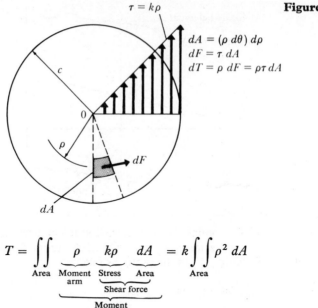

Figure 3-4

$\tau = k\rho$

$$dA = (\rho\, d\theta)\, d\rho$$
$$dF = \tau\, dA$$
$$dT = \rho\, dF = \rho\tau\, dA$$

$$T = \underbrace{\iint\limits_{\text{Area}} \underbrace{\rho}_{\substack{\text{Moment}\\\text{arm}}} \underbrace{\underbrace{k\rho}_{\text{Stress}} \underbrace{dA}_{\text{Area}}}_{\text{Shear force}}}_{\text{Moment}} = k \iint\limits_{\text{Area}} \rho^2\, dA \tag{5}$$

You may recognize the integral on the right-hand side of Equation (5) as the polar moment of inertia of the cross-sectional area. It is frequently given the symbol J, and is evaluated for a solid circular section in Equation (6), where d is the diameter of the circular cross section.

$$J = \iint\limits_{\text{Area}} \rho^2\, dA = \int_0^{2\pi} \int_0^c \rho^2 (\rho\, d\rho\, d\theta) \tag{6}$$

$$= \frac{c^4}{4} \int_0^{2\pi} d\theta = \frac{\pi c^4}{2} = \frac{\pi d^4}{32}$$

Equations (6) and (4) now permit us to write Equation (5) in the form

$$T = kJ = \frac{\tau J}{\rho} \tag{7}$$

Equation (7) can be solved for the shearing stress τ as a function of the torque and geometry:

$$\tau = \frac{T\rho}{J} \tag{8}$$

This is the very important torsional-shear-stress equation. The shear stress varies linearly with ρ. It reaches a maximum value given by $\tau_c = Tc/J$ at the surface of the specimen where $\rho = c$.

● **CAUTION** Although Equation (8), the torsional-shear-stress equation, does not involve elastic constants, it is nevertheless valid only in the linear elastic range, since the stress was assumed to be directly proportional to the strain in the derivation of the equation.

Now turn to Section SG3-1 of the Study Guide. ■ **STOP**

R3-2   Torsional deformations

In Section R3-1, we found the shearing strain on the outer surface to be given by

$$\gamma_c = \frac{c\theta}{L} \tag{9}$$

If we assume linear elastic theory, we can substitute

$$\gamma_c = \frac{\tau_c}{G} \tag{10}$$

into Equation (9) and solve for θ as a function of τ_c. The result is

$$\theta = \frac{\tau_c L}{cG} \tag{11}$$

Substitution of the shear-stress equation, $\tau_c = Tc/J$, into Equation (11) gives

$$\theta = \frac{TL}{JG} \tag{12}$$

▶ **NOTE** This important formula relating torque and the angle of twist is subject to the same limitations and assumptions that were used in the derivation of the shear-stress equation in Section R3-1, including the restriction that the quantities T, J, and G be constant throughout the length L.

If T, J, and G are constant for each of a finite number of sections n of a circular torsion member, as in the shaft shown in Example 3 of Section R3-1, we can obtain the angle of twist from the finite summation

$$\theta = \sum_{i=1}^{n} \frac{T_i L_i}{J_i G_i} \tag{13}$$

where T_i, J_i, and G_i are constant throughout each finite length L_i. This is an algebraic summation with the signs determined by the direction of the torque in each section.

If T and/or J vary continuously over the length of a torsion member, Equation (12) does not apply. However, if we assume these quantities to be constant over an infinitesimal length, then we can write Equation (12) in the form

$$d\theta = \frac{T}{JG} \, dx$$

where $d\theta$ is the differential angle of twist over an infinitesimal length dx. The finite angle of twist for a member of length L can be written in the integral form

$$\theta = \int_0^L \frac{T}{JG} \, dx \tag{14}$$

▶ **NOTE** Equations (13) and (14) are approximations. Equation (13) does not take into account the effect of discontinuities. However, these effects are usually small in applied problems. If J is not constant, Equation (14) violates assumption (b) of Section R3-1, which states that the torsion member must have a constant cross section. In practice, however, Equation (14) gives good results for slightly tapered members.

Now turn to the study program in Section SG3-2. ■ **STOP**

R3-3 Principal stresses in torsion, relationship between elastic constants

Principal stresses in torsion

In Section R3-1, we found that the shearing strain (and stress) in an elastic circular torsion member varies linearly from the center and has its maximum value at the surface. When designing structural members, we are usually interested in maximum stresses (there may also be other design considerations, such as deformations, etc.). If we are interested in maximum stresses in a circular torsion member, we need examine only the state of stress at the surface, on which the torsional shear stress ($\tau = T\rho/J$) is a maximum.

Now let us look at an infinitesimal element on the surface of a circular elastic torsion member. The element shown in Figure 3-5(a) and (b) is subjected to a state of pure shear stress. Recall what we showed in Chapter 2: that the shear stresses are of equal magnitude on perpendicular faces of the infinitesimal element (that is, $\tau_{xy} = \tau_{yx}$). What are the principal stresses? Do we have normal stresses on planes at other orientations (i.e., at orientations other than those defined by circumferential and axial lines on the specimen)? For the answers to these questions, we need only look at the Mohr's circle shown in Figure 3-6.

An examination of this circle shows that the state of pure shear stress of Figure 3-5 gives a circle centered at the origin, and that normal stresses do in fact exist on all other planes. It is further obvious that the maximum and minimum normal stresses occur on planes at 45° to the maximum shearing-stress planes, and that they are equal in magnitude to the maximum shearing stress. Figure 3-7 presents elements showing the maximum shearing stress and its relationship to the principal stresses.

If you prefer the transformation equations to Mohr's circle, you can reach the same conclusions as follows: From Figure 3-5, we see that

$$\sigma_x = 0, \qquad \sigma_y = 0, \qquad \tau_{xy} = Tc/J$$

Figure 3-5

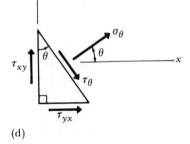

(a)

(b)   Enlarged element

(c)

(d)

Figure 3-6

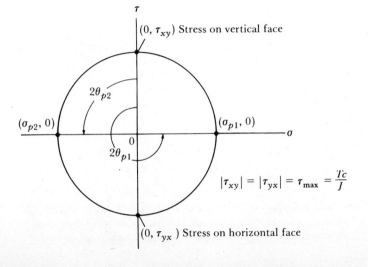

$(0, \tau_{xy})$ Stress on vertical face

$2\theta_{p2}$

$(\sigma_{p2}, 0)$

$(\sigma_{p1}, 0)$

$2\theta_{p1}$

$$|\tau_{xy}| = |\tau_{yx}| = \tau_{\max} = \frac{Tc}{J}$$

$(0, \tau_{yx})$ Stress on horizontal face

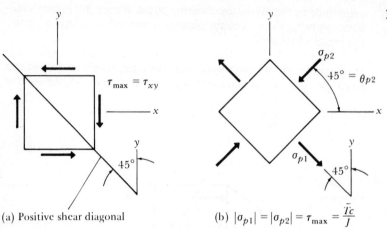

Figure 3-7

(a) Positive shear diagonal

(b) $|\sigma_{p1}| = |\sigma_{p2}| = \tau_{max} = \dfrac{Tc}{J}$

The angles locating the principal stresses and their associated planes are determined as follows:

$$\tan 2\theta_p = \frac{-2\tau_{xy}}{\sigma_x - \sigma_y} = \frac{-2(Tc/J)}{0 - 0} = \infty$$

$$2\theta_p = 90°, 270°$$

$$\theta_p = 45°, 135° \qquad (\text{or} -45°)$$

Substituting the principal angles into the normal-stress transformation equation, we obtain the principal stresses.

$$\sigma_\theta = \frac{\sigma_x + \sigma_y}{2} + \left(\frac{\sigma_x - \sigma_y}{2}\right) \cos 2\theta - \tau_{xy} \sin 2\theta$$

$$\sigma_{45°} = 0 + 0 - \left(\frac{Tc}{J}\right) \sin 90° = -\frac{Tc}{J}$$

$$\sigma_{135°} = 0 + 0 - \left(\frac{Tc}{J}\right) \sin 270° = \frac{Tc}{J}$$

Therefore

$$\sigma_{p1} = \frac{Tc}{J} \qquad \text{at} \qquad \theta = 135° \qquad (\text{or} -45°)$$

$$\sigma_{p2} = -\frac{Tc}{J} \qquad \text{at} \qquad \theta = 45°$$

These principal stresses are illustrated in Figure 3-7(b). Since the principal stresses act on planes at 45° to the planes of τ_{xy} and τ_{yx} shown in Figure 3-7(a), we conclude that $\tau_{xy} = \tau_{max}$.

We can determine the stresses σ_θ and τ_θ (Figure 3-5c) at any orientation on the surface of the torsion member by applying the theory developed in Sections R2-1 through R2-4. Similarly, we can determine the strains at any orientations according to the theory developed in Sections R2-6 and R2-7.

Relationship between elastic constants

In Section 2-9, we used, without proof, the relationship between the elastic constants, $E = 2(1 + \mu)G$. We can now prove this relationship. Consider the equivalent states of stress (a) and (b) in Figure 3-7. The maximum shearing stress for element (a) is τ_{xy}, and the maximum shearing strain is

$$\gamma_{max} = \frac{\tau_{max}}{G} = \frac{\tau_{xy}}{G} \tag{15}$$

The principal strains for element (b) of Figure 3-7 are

$$\epsilon_{p1} = \frac{1}{E}[\sigma_{p1} - \mu\sigma_{p2}] = \frac{1}{E}[\tau_{xy} - \mu(-\tau_{xy})] = \frac{\tau_{xy}(1 + \mu)}{E} \tag{16}$$

and

$$\epsilon_{p2} = \frac{1}{E}[\sigma_{p2} - \mu\sigma_{p1}] = \frac{1}{E}[-\tau_{xy} - \mu\tau_{xy}] = \frac{-\tau_{xy}(1 + \mu)}{E} \tag{17}$$

An examination of the transformation equations or Mohr's circle for strains shows that $\gamma_{max} = \epsilon_{p1} - \epsilon_{p2}$. Therefore the maximum shearing strain for element (b) is

$$\gamma_{max} = \epsilon_{p1} - \epsilon_{p2} = \frac{2\tau_{xy}(1 + \mu)}{E} \tag{18}$$

Since we have shown elements (a) and (b) of Figure 3-7 to be equivalent states of stress, it follows that the two elements must have the same principal stresses (and strains) and maximum shearing stress (and shearing strain). We can now complete our proof by equating the maximum shearing strain resulting from the state of stress on element (a), given by Equation (15), with the maximum shearing strain in element (b), given by Equation (18). The resulting equation

$$\gamma_{max} = \frac{\tau_{xy}}{G} = \frac{2\tau_{xy}(1 + \mu)}{E} \tag{19}$$

yields the relationship

$$E = 2(1 + \mu)G \tag{20}$$

Now turn to the study program in Section SG3-3. ■ **STOP**

R3-4 Power transmission

The study of torsional loading on circular members would not be complete without some consideration of power transmission. The most common application of torque-carrying circular members is power transmission. Of course, torque-carrying members are encountered in many other applications, such as structural members, but they are usually not circular in cross section, and their analysis is beyond the scope of an elementary course in mechanics of materials.

The differential work of a torque (couple) subjected to an infinitesimal displacement $d\theta$ is given by the equation

$$dW = T \, d\theta \tag{21}$$

where θ is the angle through which the torque T rotates. If the torque is constant throughout a finite rotation θ, the finite work is given by $W = T\theta$.

Power is the time rate of doing work. By definition, it follows that power P is

$$P = \frac{dW}{dt} = T\frac{d\theta}{dt} = T\omega \tag{22}$$

where ω is the angular velocity.

● **CAUTION** Watch your units. The angular velocity should be given in radians/second, unless an appropriate conversion constant is incorporated into the equation.

In the customary English system of units, power is frequently given in terms of horsepower. You will need to know the following conversion factors:

$$1 \text{ hp} = 550 \text{ lb-ft/sec} = 33{,}000 \text{ lb-ft/min} \tag{23}$$

In the International System of Units (SI), we normally express work in joules (J) and power in watts (W). You should know the following relationships:

$$1 \text{ joule (J)} = 1 \text{ newton meter (N·m)} \tag{24}$$

$$1 \text{ watt (W)} = 1 \text{ joule/second (J/s)} = 1 \text{ (N·m/s)} \tag{25}$$

You should also know the constant relating horsepower and watts:

$$1 \text{ hp} = 746 \text{ watts} \quad \text{(approximate)} \tag{26}$$

Now turn to the study program in Section SG3-4. ∎ **STOP**

Problems

3-1.1 Given the solid steel shaft loaded by torques as shown, determine the maximum shearing stress. At what section does this stress occur? ($E = 30 \times 10^6$ psi, $G = 12 \times 10^6$ psi)

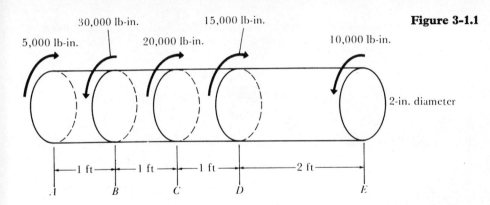

30,000 lb-in.　　　15,000 lb-in.　　　　**Figure 3-1.1**

5,000 lb-in.　　20,000 lb-in.　　　10,000 lb-in.

2-in. diameter

—1 ft—　—1 ft—　—1 ft—　—2 ft—

.1　　B　　C　　D　　E

3-1.2 In order to reduce the weight of a machine, an engineer wants to re-place a solid circular shaft 1.2 m long with a hollow circular shaft. The solid shaft is 50 mm in diameter, and the hollow shaft is to be 75 mm in outside diameter. What is the largest inside diameter that the hollow shaft may have if both shafts are to have the same max-imum shearing stress, and they are made of the same material? What is the ratio of the weight of the solid shaft to that of the hollow shaft?

3-1.3 Suppose that the two shafts of Problem 3-1.2 were made of steel, and were to have the same weight. What would the inside diameter of the hollow shaft have to be? Determine the maximum shearing stress for these two shafts and the shearing stress on the inside surface of the hollow shaft when the torque is 1750 N·m. ($E = 200$ GPa, $G = 80$ GPa)

3-1.4 Determine the maximum shearing stress in the specimen shown. ($G = 12 \times 10^6$ psi)

Figure 3-1.4

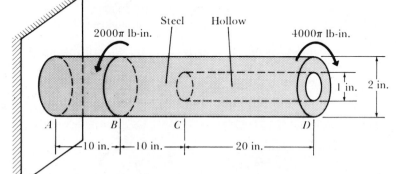

Steel　　Hollow

2000π lb-in.　　　　　　　4000π lb-in.

1 in.　2 in.

A　　B　　C　　D

—10 in.—　—10 in.—　　—20 in.—

3-1.5 Determine the percentage reduction in torque-carrying capacity of a solid circular shaft when one-half of the cross section is removed by drilling an axial hole through the length of the specimen. Note that this corresponds to a 50% reduction in weight.

3-1.6 Determine the maximum shearing stress in the stepped shaft shown, made of aluminum and steel. ($G_{st} = 70$ GPa, $G_{al} = 26$ GPa) State where this stress occurs.

Figure 3-1.6

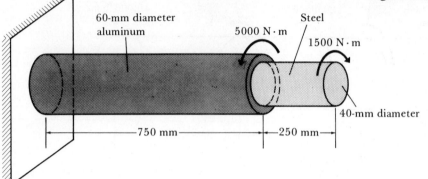

60-mm diameter
aluminum

5000 N · m

Steel

1500 N · m

40-mm diameter

—750 mm— —250 mm—

3-1.7 Determine the maximum torque that can be transmitted by the coupling shown when the shearing stress in the bolts is not to exceed
an average value of 6000 psi. Also determine the minimum diameter
of the shaft when the maximum shear stress in the shaft is 6000 psi.

1-in. diameter bolts

Figure 3-1.7

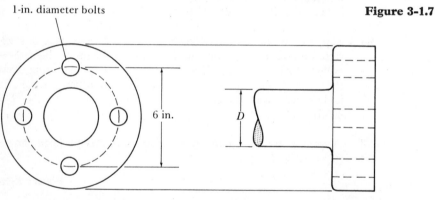

6 in.

D

Cross section   Side view (one-half of coupling)

3-1.8 Determine the maximum shearing stress in the steel shaft shown.
($E = 200$ GPa, $G = 80$ GPa, $\mu = \frac{1}{4}$)

Figure 3-1.8

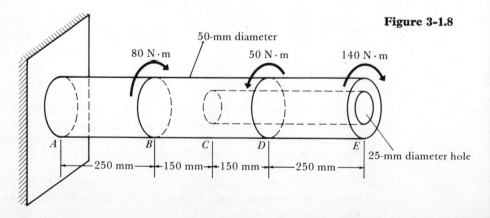

50-mm diameter

80 N · m   50 N · m   140 N · m

A B C D E

25-mm diameter hole

—250 mm— —150 mm— —150 mm— —250 mm—

3-2.1 What minimum length must a 0.1-in.-diameter aluminum wire be if it is to be twisted through one complete revolution without the shearing stress exceeding 8000 psi? What torque is required? ($G = 4 \times 10^6$ psi)

3-2.2 Determine the relative angle of twist of end E with respect to end A in Problem 3-1.1.

3-2.3 Determine the relative angle of twist of end D with respect to end A in Problem 3-1.4.

3-2.4 Determine the angle of twist of the free end relative to the fixed end in the shaft of Problem 3-1.6.

3-2.5 Determine the angle of twist of section E relative to section A in the shaft of Problem 3-1.8.

3-2.6 Determine the minimum allowable diameter of section AB of the stepped shaft shown. The shearing stress is not to exceed 42 MPa, and the angle of twist of end C relative to A is not to exceed 0.05 rad. ($E = 200$ GPa, $G = 80$ GPa)

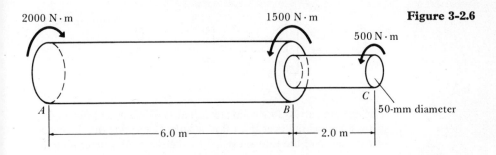

2000 N·m   1500 N·m   **Figure 3-2.6**

500 N·m

C

A B 50-mm diameter

—6.0 m—   —2.0 m—

3-2.7 Determine the minimum diameter of an 81-in.-long solid circular shaft when the shear stress is not to exceed 5000 psi and the angle of twist is not to exceed 0.008 rad. The shaft is to carry a torque of $20,000\pi$ lb-in. ($G = 5 \times 10^6$ psi)

3-2.8 A solid circular steel shaft 3 m long carries a torque of 2800 N·m. Determine the minimum allowable diameter when the shearing stress must not exceed 105 MPa and the angle of twist must not exceed 0.10 rad. ($G = 80$ GPa)

3-2.9 A torque T is applied at the top end of a vertical drill rod of length L. The rod has a constant torsional rigidity JG, and the resisting torque is assumed to be a constant (t torque/length) distributed torque along the length of the drill rod. Determine the angular deformation of the top end relative to the bottom end in terms of T, L, J, and G.

3-3.1 The motor in the sketch delivers 20,000 lb-ft of torque to the three gears through a 4-in.-diameter shaft. Each gear powers pieces of

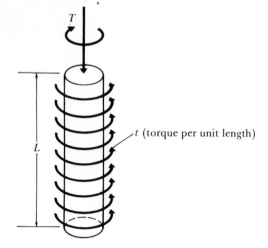

Figure 3-2.9

machinery. Gear A requires 6000 lb-ft of torque, and gear B requires 4000 lb-ft of torque. Determine the maximum tensile stress developed in the shaft.

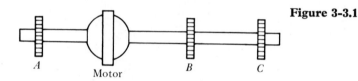

Figure 3-3.1

3-3.2 The shearing stress in a solid steel shaft is 63 MPa, and the angle of twist is 0.03 rad over the 1.0-m length. Determine the diameter of the shaft and the torque carried by the shaft. Also determine the normal and shearing stresses on an inclined plane having a slope of $\frac{3}{4}$ relative to the axis of the specimen, as shown. ($G = 80$ GPa)

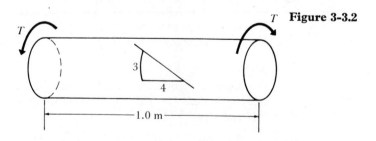

Figure 3-3.2

3-3.3 A hollow circular shaft is to carry a torque of 6000 lb-in., and the inside diameter is to be three-quarters of the outside diameter. The tensile stress is not to exceed 18,000 psi, and the shear stress is not to exceed 12,000 psi. What is the minimum diameter the shaft must have?

3-3.4 The cylindrical tube shown is to be used as a torsion member. Determine the maximum allowable torque that the tube can carry when the compressive stress in the shell is not to exceed 84 MPa. Also determine the normal and shearing stresses in the weld when the torque is at a maximum. Assume that the thickness of the weld is the same as the tube thickness. ($G = 80$ GPa, $\mu = \frac{1}{4}$).

Figure 3-3.4

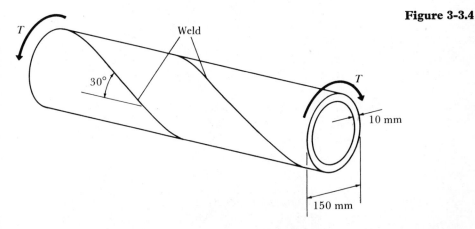

3-3.5 What strain should the strain gage indicate when the torque is applied to the specimen shown? Is the strain tensile or compressive? ($E = 200$ GPa, $\mu = \frac{1}{4}$)

Figure 3-3.5

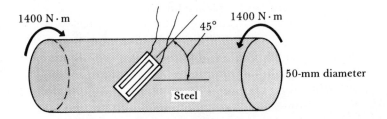

3-3.6 Determine the strain at 30° to the axis of the 2-in.-diameter torsion specimen shown when the torque is 8000 lb-in. Is the strain tensile or compressive?

Figure 3-3.6

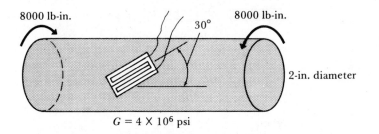

3-3.7 A delta strain rosette is applied to a torsion specimen, as shown. When the strains are

$$\epsilon_a = 0, \qquad \epsilon_b = 150 \times 10^{-6} \text{ in./in.}, \qquad \epsilon_c = -150 \times 10^{-6} \text{ in./in.},$$

determine the magnitude and direction of the torque applied to the right-hand end of the shaft. ($G = 12 \times 10^6$ psi)

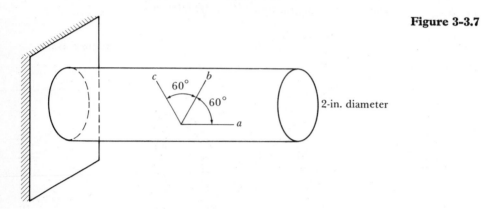

Figure 3-3.7

3-4.1 Determine the minimum diameter required of a solid circular shaft to transmit 10 hp at 360 rpm, given that the maximum shearing stress is not to exceed 8000 psi.

3-4.2 An electric motor drives the line shaft, as shown. The angular velocity of the line shaft is 65 rad/s. The shaft supplies 20 kW to equipment driven by the gear C, and 60 kW to equipment driven by pulley A. The maximum tensile stress in the line shaft is not to exceed 40 MPa, and the maximum shearing stress is not to exceed 50 MPa. Determine the minimum diameter that the solid steel shaft may have. ($G = 80$ GPa)

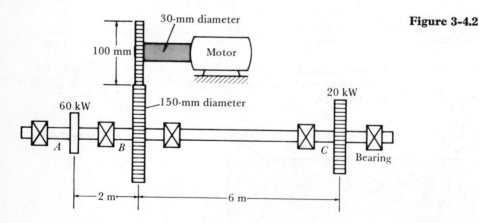

Figure 3-4.2

3-4.3 Determine the maximum shearing stress in the 30-mm-diameter motor shaft of Problem 3-4.2.

3-4.4 Twenty-five horsepower is supplied at 300 rpm at pulley A. A steel shaft 1 in. in diameter transmits this power to pulleys B, C, and D. Pulleys B and D each supply 10 hp to drive equipment, and pulley

C supplies 5 hp to power a saw. Determine the maximum shearing stress in section *BC* of the shaft.

Figure 3-4.4

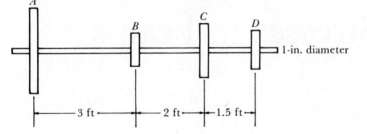

Figure 3-4.4

3-4.5 A solid circular aluminum shaft 3 in. in diameter and 5 ft long rotates at 200 rpm. If the shear stress is not to exceed 8000 psi and the angle of twist is not to exceed 0.1 rad, what maximum horsepower can the shaft deliver? ($G = 4 \times 10^6$ psi)

3-4.6 A hollow aluminum shaft is to have an inside diameter equal to three-fourths of the outside diameter. The shaft is to transmit 1.5 MW at an angular velocity of 190 rad/s. Determine the minimum allowable outside diameter, given that the shearing stress is not to exceed 105 MPa and the angle of twist is not to exceed 0.07 rad/m. ($G = 26$ GPa)

3-4.7 Determine the minimum allowable diameter of a solid circular steel shaft which must carry 350 hp at 200 rpm. The shearing stress is not to exceed 8000 psi, and the angle of twist is not to be greater than 0.01 rad/ft of length. ($G = 12 \times 10^6$ psi)

3-4.8 Determine the minimum permissible diameter for a solid circular steel shaft that is to transmit 2.2 MW at 160 rad/s. The allowable shearing stress is 84 MPa, and the angle of twist is not to exceed 0.05 rad. The shaft is 5 m long. ($G = 80$ GPa)

Chapter four

Stresses in beams

R4-1* Beams: Internal reactions, shear and bending-moment diagrams and equations

In a statics course you learn how to determine external reactions, as well as reactions created by internal connections holding the various members of a structure together. In this section, we shall learn how to determine the internal reactions that hold the various parts of a given member together.

Before we get into a detailed discussion of this topic, however, it may help to quickly review the vocabulary that is commonly used to describe methods of supporting beams. The following terms refer to methods of support (or constraint), not to the type of loading.

Simple beam A beam supported by a hinge pin at one end and a roller at the other.

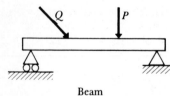

Beam

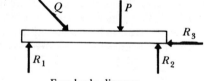

Free-body diagram

Cantilever beam A beam supported at one end only, to prevent translation as well as rotation of that end.

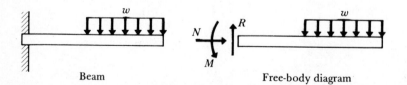

Beam Free-body diagram

* Large portions of Section 4-1 are excerpted, with the permission of Addison-Wesley Publishing Company, from K. C. Muhlbauer, *Statics: An Individualized Approach*, Addison-Wesley, Reading, Mass., 1972.

Overhanging beam A beam supported by a hinge pin and a roller, with either or both ends extending beyond the supports.

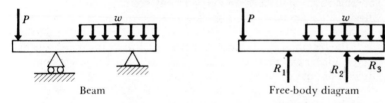

Beam   Free-body diagram

We are now ready to determine the internal reactions in a beam. Figure 4-1(a) shows a simply supported beam, and Figure 4-1(b) shows a free-body diagram (in perspective) of that portion of the beam which lies to the left of section *aa* together with the forces that the beam must produce for equilibrium. The reactive forces that the beam must produce are shown on three different views of the cross-sectional area of the beam, but they are of course actually produced simultaneously on just one section. The resultant of the normal force N acts through the centroid of the cross-sectional area *abcd* and the couple M is determined with respect to the centroidal z axis of this cross-sectional area. The cross section *abcd* is shown as a rectangle for convenience only. It may have any arbitrary shape.

Figure 4-1

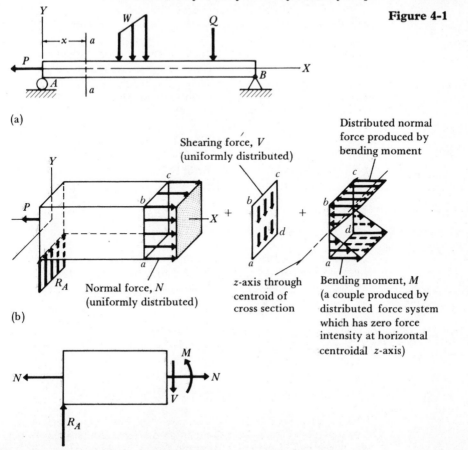

(a)

Shearing force, V
(uniformly distributed)

Distributed normal force produced by bending moment

Normal force, N
(uniformly distributed)

z-axis through centroid of cross section

Bending moment, M (a couple produced by distributed force system which has zero force intensity at horizontal centroidal z-axis)

(b)

(c)

It is very tedious to draw free-body diagrams in perspective, as shown in Figure 4-1(b), because, in order to get an accurate picture of the internal reactions at any section along the beam, we would have to investigate many sections. This is why we shall next develop methods to enable us to draw continuous shear and bending-moment diagrams over the length of the beam. With the aid of these diagrams, the design engineer can readily determine the necessary cross-sectional dimensions at any point along the beam. The construction of the shear and bending-moment diagrams is greatly facilitated by the establishment of the relations that exist among load, shear, and bending moment. It is seldom necessary to draw a "normal-force" diagram, since the variation of N along the length of the beam can usually be determined by examining only one or two separate sections.

Figure 4-2(a) shows a simply supported beam subjected to a distributed force system. Figure 4-2(b) shows a free-body diagram of a small portion of the beam between sections C and D. Only the resultants of the shear force V and of the couple M are shown, but it is important that you understand the true nature of these reactions, as shown in Figure 4-1(b). The normal force N is not shown on this free-body diagram, since it is zero for this loading condition. *The reactions V and M as shown in this diagram are positive according to the standard sign convention for beams.* Positive directions are defined as follows.

Shear is considered *positive* if the left-hand portion of the beam has a tendency to move up after a section is cut. *Bending moment* is considered positive if it tends to deform the beam in such a way that the top fibers are in compression and the bottom fibers are in tension, as shown in Figures 4-1(b) and 4-2(b).

In Figure 4-2, assume that Δx is sufficiently small so that the variation of the load q from C to D can be neglected. The relationship between the load and the shear can be established by summing forces in the vertical direction on the free-body diagram of Figure 4-2(b):

$$V - (V + \Delta V) + q\,\Delta x = 0$$
$$\Delta V = q\,\Delta x$$

Dividing both sides of the equation by Δx and then letting x approach zero, we have

$$\frac{dV}{dx} = q \tag{1}$$

To establish the relationship between shear and bending moment, we take the sum of the moments about D:

$$(M + \Delta M) - M - (V\,\Delta x) - \left(q\,\Delta x\,\frac{\Delta x}{2}\right) = 0$$

$$\Delta M = V\,\Delta x + \tfrac{1}{2}q(\Delta x)^2$$

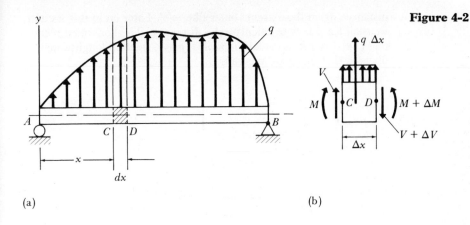

Figure 4-2

(a)   (b)

Dividing both sides of the equation by x and then letting x approach zero, we obtain

$$\frac{dM}{dx} = V \tag{2}$$

Now turn to the Study Guide to learn how these equations can be applied to construct shear and bending-moment diagrams, and how to write shear and bending-moment equations. ■ **STOP**

R4-2 Flexure stresses in beams

The bending or flexing of a beam, as shown in Figure 4-4, creates normal stresses along the axis of the beam. Look at lines aa' and bb' in Figures 4-3 and 4-4. In Figure 4-3, they are parallel, and in Figure 4-4 they are not parallel. You can see that line ab in Figure 4-4 is shorter than line ab in Figure 4-3; it has been compressed. You can also see that line $a'b'$ in Figure 4-4 is longer than line $a'b'$ in Figure 4-3; it has been elongated, and is in tension. Since line ab is compressed and line $a'b'$ is elongated, there must be some line in between which is neither compressed nor elongated. We shall assume this line to be represented by $a''b''$, shown in Figure 4-4. Since the length of line $a''b''$ in Figure 4-4 remains unchanged, this line is not strained, and we call it the *neutral fiber*. This neutral fiber $a''b''$ is shown at a

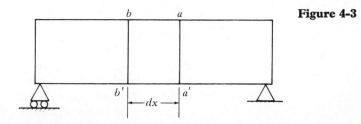

Figure 4-3

distance c from the extreme outer fiber $a'b'$. Later on in this section, we shall learn how to calculate the distance c. The horizontal plane that contains $a''b''$ is called the *neutral plane*, since all longitudinal fibers in that plane are unstrained.

Figure 4-4

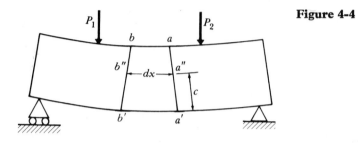

Now you have a general idea of how bending creates normal strains (and stresses) of tension and compression along the axis of the beam. The associated stresses are commonly called *flexure stresses*. Let's expand this idea into an applicable formula that may be used to obtain quantitative results.

● **CAUTION** As we expand the concept of beam flexure into a usable formula, called the *flexure formula*, we shall face certain limitations in the theory. Memorize these limiting conditions and be ready to explain them.

▶ **NOTE** The first limiting condition is that the cross section of the beam must have an axis of symmetry in the plane of the loads, such as those in the examples of Figure 4-5.

Figure 4-5

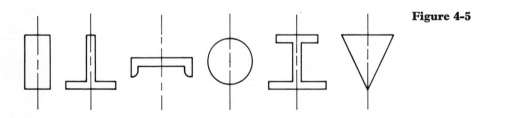

If the plane of loading is parallel to, but not coincident with, the axis of symmetry of the cross section, a torsional load is created and torsional stresses exist. We are not prepared to analyze torsional members having noncircular cross sections. If the cross section is unsymmetric, two things will happen: (1) The beam will twist, creating torsional stresses, unless the plane of loading passes through a point called the *shear center* of the cross section. We shall not investigate the concept of shear center in this text. (2) The neutral plane will not remain horizontal, which creates complications in the development of the flexure theory that are beyond the scope of this text.

The first part of the derivation of the flexure formula consists of establishing the relationship between the geometry of deformation and strain. The second part consists of the application of the principle of equilibrium. To begin the first part, let's take the section of the beam between aa' and bb' in Figure 4-2 and show it on a larger scale (Figure 4-6).

▶ **NOTE** We have presented the second limitation. The lines aa' and bb' were shown as straight lines in Figure 4-4, which implies that the cross sections containing these lines remain plane after the beam bends. For a cross section to remain plane after bending, it must have no shear force acting on it. A more rigorous solution taken from the mathematical theory of elasticity shows that cross-sectional warpage occurs if the cross section carries a shear force in addition to a bending moment. However, this warpage has such a small effect that the assumption of cross sections remaining plane is considered valid even when shear forces are present.

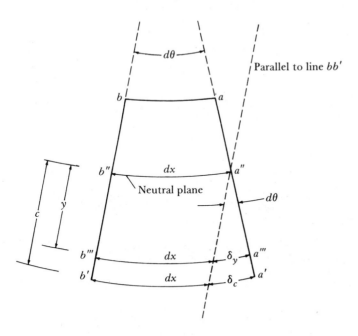

Figure 4-6

According to the definition of normal strain in Chapter 1, the expressions for the strains of fibers $a'b'$, $a''b''$, and $a'''b'''$, which we shall call ϵ_c, ϵ_0, and ϵ, respectively, are

$$\epsilon_c = \frac{\delta_c}{dx} \qquad (3a)$$

$$\epsilon_0 = \frac{0}{dx} = 0 \quad \text{(a neutral fiber)} \tag{3b}$$

$$\epsilon = \frac{\delta_y}{dx} \tag{3c}$$

Remember, strain is deformation per unit length. The quantity ϵ is the strain of an arbitrary fiber located a distance y from the neutral fiber, as shown in Figure 4-6.

▶ **NOTE** For equations (3a), (3b), and (3c) to be valid, we must live with a third limiting condition. Do you agree that, according to these three equations, the three fibers in question must have the same original length dx? Good. For all axial fibers of a beam between two cross sections, such as aa' and bb', to have the same length before bending takes place, the beam must be originally straight, and it must have a constant cross section.

From geometry, we see that $\delta_c = c\,d\theta$ and $\delta_y = y\,d\theta$, and if we substitute these expressions into Equations (3a) and (3c), we obtain

$$\epsilon_c = \frac{c\,d\theta}{dx} \tag{4a}$$

$$\epsilon = \frac{y\,d\theta}{dx} \tag{4b}$$

If we divide equation (4b) by equation (4a), we obtain

$$\epsilon = \frac{y}{c}\,\epsilon_c \tag{5}$$

▶ **NOTE** Let us now introduce a fourth condition, that Hooke's law must apply to the individual fibers in the axial direction. This condition implies that stresses in the axial direction are restricted to magnitudes within the proportional limit of the material.

We may write the equations for the normal stresses of fibers $a'b'$ and $a''b''$, respectively, as

$$\sigma_c = E_c\epsilon_c \quad \text{and} \quad \sigma = E\epsilon \tag{6}$$

where E_c and E are the moduli of elasticity of fibers $a'b'$ and $a''b''$, respectively.

▶ **NOTE** A fifth and final limiting condition is that the modulus of elasticity must be the same for all fibers, and it must be the same in tension and compression.

Equations (6) now become

$$\sigma_c = E\epsilon_c \quad \text{and} \quad \sigma = E\epsilon \tag{7}$$

If we substitute the expressions for strain, given by Equations (7), into Equation (5), we obtain

$$\frac{\sigma}{E} = \frac{y}{c}\frac{\sigma_c}{E}$$

$$\sigma = \frac{y}{c}\sigma_c \tag{8}$$

Equation (8) implies that the stress in any axial fiber is proportional to the distance y from the neutral plane.

To complete the derivation of the flexure-stress formula, we must consider a free-body diagram of the small section of the beam in Figure 4-4 between lines aa' and bb'. This free-body diagram is shown in Figure 4-7. In Figure 4-7, the bending moment on section aa' is expressed as a stress distribution. The stress is shown as compression at the top and tension at the bottom, because the moment is "pushing" on the cross section at the top and "pulling" at the bottom. Equation (8) tells us that the stress is distributed linearly. The stress is zero at $a''b''$, because it is in the neutral plane (Equation 3b).

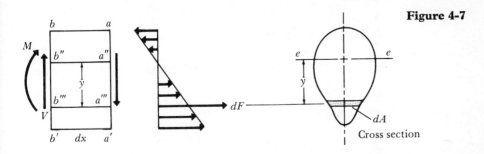

Figure 4-7

If we assume the beam to be in equilibrium, then any part of the beam, such as that shown in Figure 4-7, must also be in equilibrium The summation of forces in the axial direction gives

$$\sum F = 0$$

$$\int_A dF = 0 \qquad \text{where } A = \text{cross-sectional area}$$

$$\int_A \sigma_c \frac{y}{c}\, dA = 0 \qquad \text{since } dF = \sigma\, dA \text{ and } \sigma = \frac{y}{c}\sigma_c, \text{ from Equation (8)}.$$

$$\frac{\sigma_c}{c}\int_A y\, dA = 0 \qquad \text{so that } \int_A y\, dA = 0, \text{ since } \sigma_c \text{ and } c \text{ are constants}.$$

If you do not know what $\int_A y\, dA$ is, briefly review the subject of centroids of areas as discussed in your statics text. ■ **STOP**

If you know that

$$\int_A y\, dA = \bar{y}A$$

where $\bar{y}$ is the distance from the axis from which y is measured to the centroid of the cross section, excellent. You are correct. Now we have

$$\bar{y}A = 0$$

or

$$\bar{y} = 0 \qquad \text{since } A \text{ cannot be zero} \tag{9}$$

What does Equation (9) mean? Since y is measured from the neutral plane, we conclude that the distance in the y direction from the neutral plane to the centroid of the cross section is zero. Thus the centroid is on the neutral plane or *neutral axis* shown by line *ee* in Figure 4-7. To locate the neutral plane or neutral axis, we merely have to locate the centroid of the cross section.

If we sum moments about the neutral or centroidal axis (*ee*) or about point a'' in Figure 4-7, we obtain

$$\sum M = 0$$

$$\int_A y \, dF - M - V \, dx = 0$$

A review of Section R4-1 shows the relationship between V and M to be

$$\frac{dM}{dx} = V$$

or

$$V \, dx = dM$$

Therefore we have

$$\int_A y \, dF - M - dM = 0$$

Since dM is infinitesimal, it is negligible compared with other terms, and we have

$$\int_A y \, dF - M = 0$$

$$M = \int_A y \sigma \, dA \qquad \text{since } dF = \sigma \, dA.$$

$$M = \int_A y \frac{y}{c} \sigma_c \, dA \qquad \text{since } \sigma = \frac{y}{c} \sigma_c, \text{ from Equation (8).}$$

$$M = \frac{\sigma_c}{c} \int_A y^2 \, dA \qquad \text{since } \sigma_c \text{ and } c \text{ are constants.}$$

If you do not know what $\int_A y^2 \, dA$ is, briefly review the subject of moments of inertia of areas as discussed in your statics text. ■ **STOP**

If you know that $\int_A y^2 \, dA$ is the moment of inertia of the cross-sectional area about *ee* (the neutral or centroidal axis), you are correct.

Thus

$$M = \frac{\sigma_c I}{c} \quad \text{where } I = \int_A y^2 \, dA$$

or

$$M = \frac{\sigma I}{y} \quad \text{since } \sigma_c = \sigma \frac{c}{y} \text{ from Equation (8)}$$

or

$$\sigma = \frac{My}{I} \tag{10}$$

which is the flexure-stress formula. Do you remember all the limiting conditions? Let's list them here.

1 The beam cross section must have an axis of symmetry parallel to the loading direction.
2 Cross sections must remain plane after bending. Strictly speaking, this requires that the cross section be free of a shear force. However, since the shear force usually causes only slight warpage of the cross section of relatively long beams, the requirement is not usually enforced unless the beam is short and deep.
3 The beam must be straight with a constant cross section.
4 Hooke's law must apply to the individual longitudinal fibers. For the flexure formula to be valid, the flexure stress must not exceed the proportional limit of the material.
5 The modulus of elasticity must be the same for all longitudinal fibers, and must be the same in tension and compression.

Take a break and relax if you want. Then turn to Section SG4-2 of the Study Guide, where you will find more explanation and some examples. ■ **STOP**

R4-3 Horizontal and vertical shear stresses in beams

Since you should now understand that internal bending moments cause normal stresses in beams, you may think it reasonable to ask whether the shear forces also cause stresses in beams. The answer is yes; and if you recall that any force that acts parallel to an area causes a shear stress on that area, you will recognize the obvious: that shear forces in beams do indeed cause shear stresses. If these shear stresses are uniformly distributed over the entire section, as was assumed in Chapter 1, their value will be

$$\tau = \frac{V}{A}$$

where:

τ = average shear stress
V = shear force in the beam
A = cross-sectional area on which V acts

However, we cannot assume the shear stresses in beams to be uniform. Thus we must find an applicable expression for calculating these shear stresses.

Figure 4-8 shows a sequence of free-body diagrams of the short length of beam between the cross sections aa' and bb' of Figure 4-4 in Section R4-2. Figure 4-8(b) shows the shear forces and bending moments with which you should be thoroughly familiar by now. Figure 4-8(c) has the bending moments replaced by their normal-stress distributions, which was described in Section R4-2. Figure 4-8(d) is a free-body diagram of just the lower part of the beam element, and in Figure 4-8(e) the normal-stress distributions are replaced by their resultant normal forces. Figure 4-8(f) shows the beam cross section. You should be familiar with every force in Figure 4-8(c) and (d), except the shear force V_H, which we will discuss now.

You recall that bending moments are usually not constant; they vary from one section to another along the beam. Therefore we should assume that M_1 and M_2 are unequal. We shall assume here that M_2 is greater than M_1; thus F_2 is larger than F_1. Now we realize that V_H must exist for our beam to be in equilibrium. The equilibrium equation is (refer to Figure 4-8e)

$$\sum F = 0 = F_2 - F_1 - V_H \tag{11}$$

(If we assume M_1 to be greater than M_2, then V_H will be in the opposite direction, as shown.) If we express the normal forces F_1 and F_2 in terms of stress and solve for V_H, we find that Equation (11) becomes

$$V_H = \int_{A_F} \sigma_2 \, dA - \int_{A_F} \sigma_1 \, dA \tag{12}$$

where A_F is the area of the cross section between h and c. Since

$$\sigma_1 = \frac{M_1 y}{I} \quad \text{and} \quad \sigma_2 = \frac{M_2 y}{I}$$

Figure 4-8

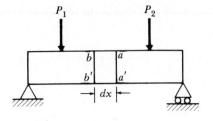

(a)

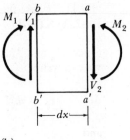

(b)

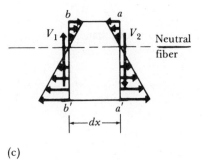

Neutral
fiber

(c)

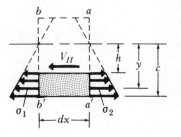

(d)

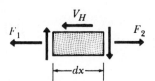

(e)

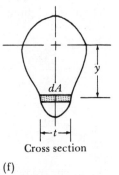

Cross section

(f)

Equation (12) becomes

$$V_H = \int_{A_F} \frac{M_2 y}{I}\, dA - \int_{A_F} \frac{M_1 y}{I}\, dA \qquad (13)$$

$$V_H = \frac{M_2 - M_1}{I} \int_{A_F} y\, dA$$

▶ **NOTE** We introduced the flexure formula into our derivation, thus also introducing the limiting conditions of the flexure formula.

Quickly review these limiting conditions listed at the end of Section R4-2. ■ **STOP**

Since M_2 and M_1 act on cross sections separated by the distance dx, we can write

$$M_2 - M_1 = dM$$

and

$$V_H = \frac{dM}{I} \int_{A_F} y \, dA \tag{14}$$

According to our definition of uniform stress, stated in Chapter 1, we have

Force = (stress)(area)

or

$$V_H = (\tau)(t \, dx) \tag{15}$$

where the stress τ is assumed to be uniformly distributed on the infinitesimal area $t \, dx$. It is valid to assume τ to be uniform over the infinitesimal dimension dx and over the dimension t, provided that t is small:

$$\tau t \, dx = \frac{dM}{I} \int_{A_F} y \, dA \tag{16}$$

$$\tau = \frac{dM}{dx} \frac{1}{It} \int_{A_F} y \, dA$$

From our study of shear and moment diagrams, we know that

$$V = \frac{dM}{dx}$$

Therefore Equation (16) may be written as

$$\tau = \frac{V}{It} \int_{A_F} y \, dA \tag{17}$$

where V is either V_1 or V_2 since they differ by only an infinitesimal amount. The integral in Equation (17) is the statical (or first) moment of the area A_F about the neutral (or centroidal) axis. If we designate this integral by the symbol Q, the final expression for the shear is

$$\tau = \frac{VQ}{It} \tag{18}$$

Remember, for the shear stress τ to exist, the internal bending moment must be nonuniform. If you do not completely understand this, go through the derivation again until it is clear.

▶ **NOTE** If we assume Equation (15) as well as Equation (18) to be exact, the stress must be uniform over the thickness t. However, more rigorous theory shows that τ is not uniform. Thus we must assume it to be the average stress across the thickness t. For a rectangular section of depth twice the width, the maximum stress is about 3% greater than the average. For a square beam, the maximum stress is about 12% greater than the average. And if the width is four times the depth, the difference is almost 100%. We conclude that Equation (18) gives inaccurate results for stresses in beams where the width is large compared with the depth, such as the flange of an I-beam. However, Equation (18) gives good results where the width is small compared with the depth, such as in the web of an I-beam. Also the variation of stress across the dimension t is greater if the sides of the beam are not parallel.

If you go back to Chapter 2, you will find that the shear stresses on mutually perpendicular planes at any point on a stressed member have the same magnitude. Thus we can say that the shear stresses on the horizontal (longitudinal) and vertical (transverse) planes at the same point on a beam must have the same magnitude. This fact is illustrated in Figure 4-9 (see page 98).

Figure 4-9(a) shows the cross section of a beam on which a shear force V is acting. Figure 4-9(b) illustrates the horizontal and vertical shear stresses at a specific location, and Figure 4-9(c) is an enlarged view of the differential element in Figure 4-9(b). [The normal stresses that must be present are omitted for clarity from Figure 4-9(b)].

Before turning to the Study Guide, we should consider an additional concept about shear stress, without going into a lengthy derivation. The shear force V_H in Figure 4-8(d) and (e) is on a horizontal plane. This shear force can also act on a vertical plane parallel to the plane of symmetry of the cross section, in beams with flanges, as shown in Figure 4-10 on page 99.

Figure 4-10 shows a sequence of diagrams similar to those in Figure 4-8. Figure 4-10(b) shows the shear forces and bending moments on a differential length of the beam. Figure 4-10(c) shows a small portion of the flange. The forces F_2 and F_1 in Figure 4-10(c) result from the normal stresses acting on the shaded area produced by the actions of the bending moments. If we assume that M_2 is larger than M_1, then F_2 is larger than F_1, and the shear force V_H must exist on the vertical surface in Figure 4-10(c) for the beam to be in equilibrium. From here, the derivation is identical to the previous derivation in this section, and the shear stresses τ shown in Figure 4-10(d), created by the shear force V_V, are determined by Equation (18):

$$\tau = \frac{V}{It}\int_{A_F} y\, dA = \frac{VQ}{It}$$

Figure 4-9

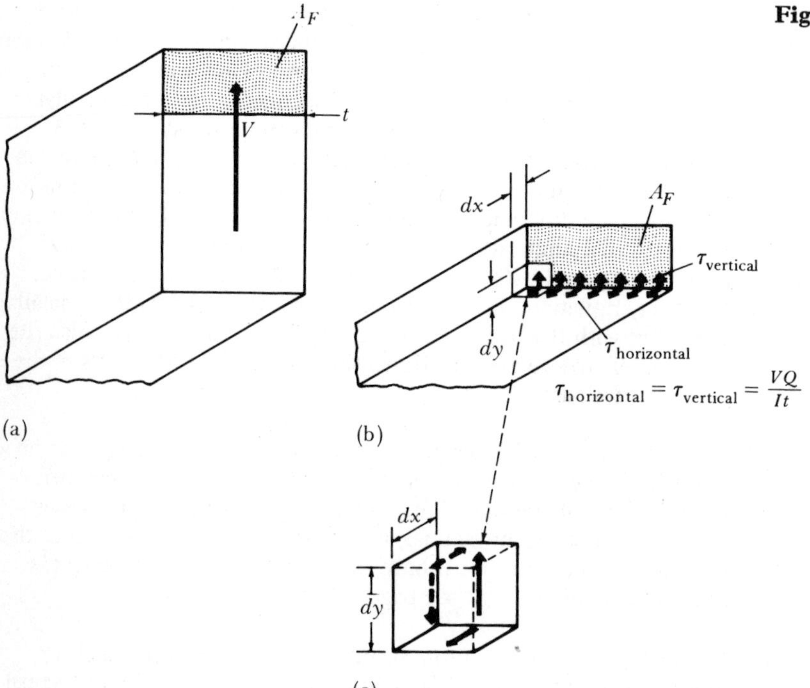
A_F
t
V
dx
A_F
$\tau_{vertical}$
dy
$\tau_{horizontal}$
$\tau_{horizontal} = \tau_{vertical} = \frac{VQ}{It}$
dx
dy

(a) (b) (c)

Do you remember all the limiting conditions of Equation (18)? Let's list them before we turn to the Study Guide.

1–5 The same as those of the flexure formula listed at the end of Section R4-2.

6 The stress is assumed to be uniform over the thickness t. Thus t should be small compared with other dimensions, and the sides of the beam should be parallel.

Examples of the quantity t are shown in Figure 4-11 on page 99. In each of these cases, we are interested in the shear stress on plane AA.

Now turn to Section SG4-3 in the Study Guide for some examples and more explanation. ■ **STOP**

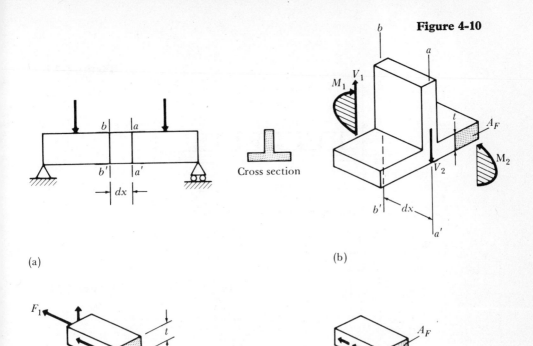

Figure 4-10

(a)

Cross section

(b)

(c)

(d)

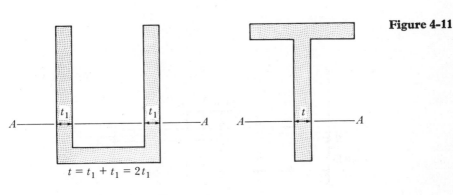

Figure 4-11

$t = t_1 + t_1 = 2t_1$

Cross sections

Problems

For each of Problems 4-1.1 through 4-1.12, draw the shear and bending-moment diagrams and write the shear and bending-moment equations for the section between points *A* and *B*. (Label *all* points of change.)

4-1.1

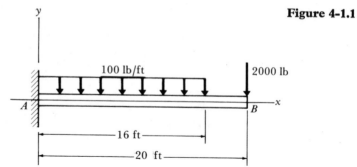

Figure 4-1.1

4-1.2

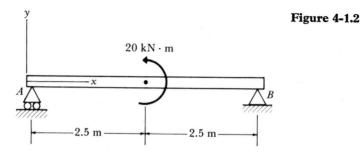

Figure 4-1.2

4-1.3

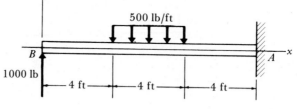

Figure 4-1.3

4-1.4

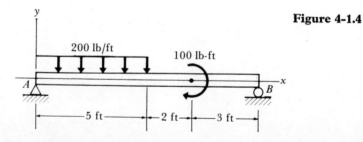

Figure 4-1.4

4-1.5

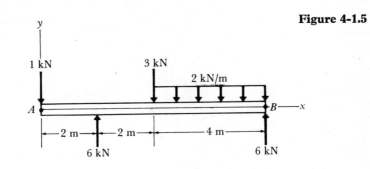

Figure 4-1.5

4-1.6

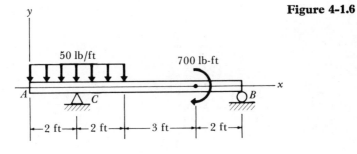

Figure 4-1.6

4-1.7

Figure 4-1.7

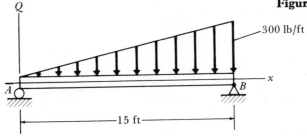

4-1.8

Figure 4-1.8

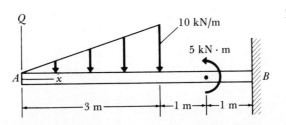

4-1.9

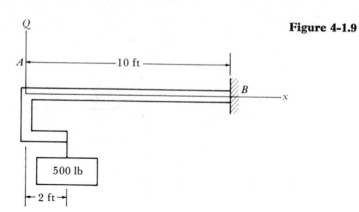

Figure 4-1.9

4-1.10

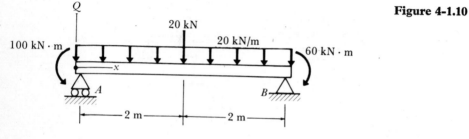

Figure 4-1.10

4-1.11

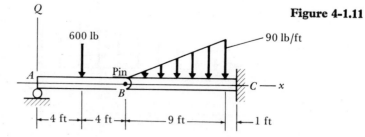

Figure 4-1.11

4-1.12

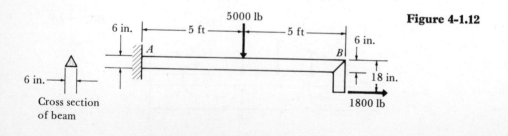

Figure 4-1.12

4-2.1 A cantilever beam 2 in. wide, 8 in. high, and 12 ft long carries a load that varies uniformly from zero at the free end to 120 lb/ft at the wall. Compute the flexure stress in a fiber 3 in. below the neutral axis in a section 9 ft from the free end.

4-2.2 The beam has a cross section with width b millimeters and depth $4b$ millimeters. The flexural stress is not to exceed 84×10^6 Pa = 84 MPa. Determine the dimension b.

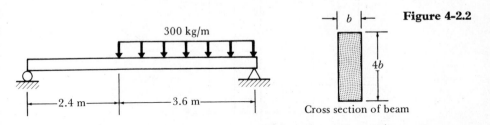

300 kg/m

2.4 m 3.6 m

Figure 4-2.2

b

$4b$

Cross section of beam

4-2.3 In the laboratory test of a beam loaded by end couples, the fibers at layer AB are found to lengthen by 0.0012 in. and the fibers at CD shorten by 0.0036 in. in the 8 in. gage length. Given that $E = 15 \times 10^6$ psi, what is the flexural stress at the top and bottom fibers?

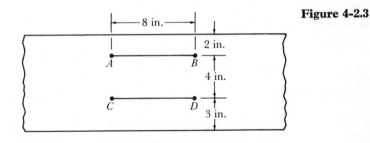

Figure 4-2.3

8 in.

2 in.

A B

4 in.

C D

3 in.

4-2.4 The maximum flexure stress at a certain cross section in a rectangular solid beam 100 mm wide and 200 mm deep is 600 kPa. (a) Determine the maximum flexure stress in the beam when the dashed portion of the cross section is removed. (b) Determine the percentage increase in stress and the percentage decrease in mass caused by removing the dashed portion of the cross section. (See Figure 4-2.4 on page 104.)

4-2.5 A 15-in., 50-lb I-beam (see Appendix 2, Table 2) is simply supported at the ends. The beam in turn supports a centrally concentrated load of 12,000 lb and a uniformly distributed load of 1000 lb/ft, including the weight of the beam. Compute the maximum length of the beam so that the flexural stress will not exceed 20,000 psi.

4-2.6 Determine the maximum tensile and compressive flexure stresses developed in the overhanging beam shown (see page 104).

4-2.7 A beam carries a concentrated load W and a total uniformly distributed load of $4W$, as shown. What safe value of W can be applied

Figure 4-2.4

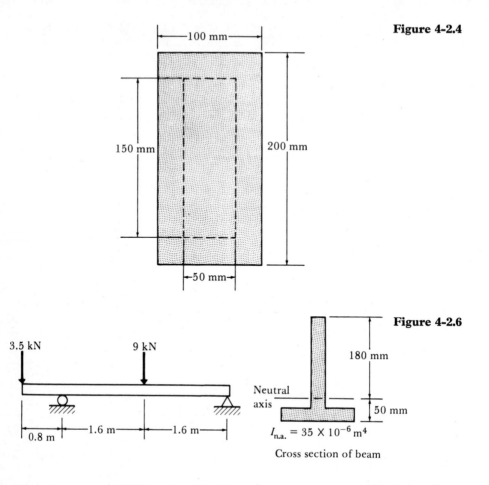

Figure 4-2.6

Neutral axis

180 mm

50 mm

$I_{n.a.} = 35 \times 10^{-6}\,m^4$

Cross section of beam

3.5 kN

9 kN

0.8 m

1.6 m

1.6 m

if the flexural compressive stress must be ≤ 14,000 psi and the flexural tensile stress must be ≤ 9000 psi? Can a greater load be applied if the section is inverted? Explain.

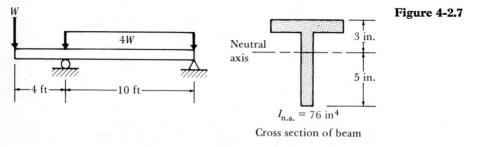

Figure 4-2.7

W

4W

4 ft

10 ft

Neutral axis

3 in.

5 in.

$I_{n.a.} = 76\,in^4$

Cross section of beam

4-2.8 A T-beam supports the three concentrated loads shown. Prove that the neutral axis is 87 mm above the bottom and that the moment of inertia about the neutral axis is $36.2 \times 10^{-6}\,m^4$. Then use these values to determine the maximum allowable value of P, given that the flexural tensile stress must be ≤ 27 MPa and the flexural compressive stress must be ≤ 50 MPa.

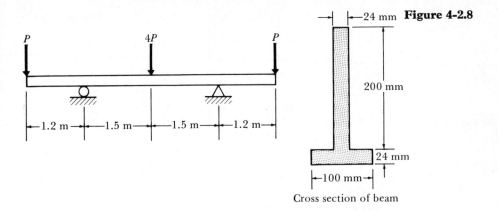

Figure 4-2.8

Cross section of beam

4-3.1 Determine the maximum permissible value for the load w (kg/m), given that the compressive flexural stress is not to exceed 6.2 MPa and the horizontal shearing stress is not to exceed 0.5 MPa.

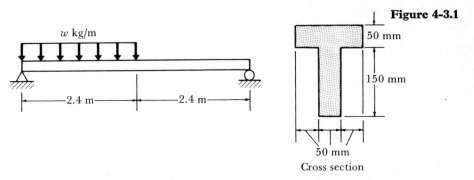

Figure 4-3.1

Cross section

4-3.2 Show that the maximum horizontal or vertical shear stress

  a) In a rectangular beam is $\frac{3}{2} V/A$,

  b) In a circular beam is $\frac{4}{3} V/A$,

  c) In a thin-walled tubular beam is $2 V/A$,

  d) In an I-beam is approximately V/A_w,

  where A is the cross-sectional area and A_w is the area of web of the I-beam.

4-3.3 A channel section carries two concentrated loads, each of magnitude P (in newtons), and a distributed load of total magnitude $6P$ (newtons). Verify that the neutral axis is 54 mm above the bottom and the moment of inertia about the neutral axis is 18×10^{-6} m⁴. Then determine the maximum value of P so that the flexure tensile stress will not exceed 28 MPa, the flexure compressive stress will not exceed 70 MPa, and the horizontal or vertical shear stress will not exceed 42 MPa.

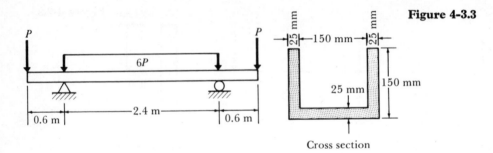

Figure 4-3.3

$6P$

2.4 m

0.6 m   0.6 m

−150 mm−

25 mm

150 mm

25 mm   25 mm

Cross section

4-3.4 A hollow wooden box beam is to be made from planks, as shown. At the section considered, the total internal vertical shear in the beam will be 930 lb, and the internal bending moment will be 50 lb-ft. Nailing is done with 16-penny nails that are good for 50 lb each in shear. What must the spacing of the nails be? (The wood planks are considered to be full size.)

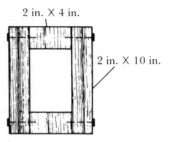

2 in. × 4 in.

2 in. × 10 in.

Figure 4-3.4

4-3.5 Two W 8 × 31 (see Appendix 2) beams are to be fastened together with two rows of rivets so as to make the two beams act as a unit. At the section considered, the total internal vertical shear is 40,000 lb, and the internal bending moment is 2700 lb-ft. If we use $\frac{3}{4}$-in.-diameter rivets, which in single shear can resist 6000 lb each, what is the proper rivet spacing?

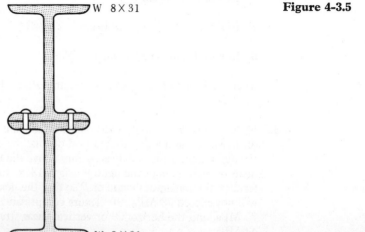

W 8 × 31

W 8 × 31

Figure 4-3.5

4-3.6 If $P = 6.7$ kN, what is the shearing stress at horizontal layers 25 mm apart from top to bottom at the section of maximum internal shear force? The neutral axis is 87.5 mm above the bottom, and the moment of inertia of the cross-sectional area above the neutral axis is 35.5×10^{-6} m^4. Verify.

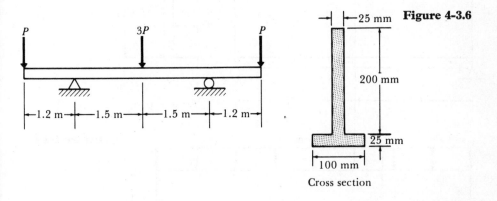

Figure 4-3.6

Cross section

4-3.7 Determine the shearing stress along section AA shown in the figure, for a W 10 × 21 beam (see Appendix 2) resisting an internal shear force of 20,000 lb and an internal bending moment of 30,000 lb-in. Given that the internal bending moment is positive and increasing, show on a sketch the direction of the shear stress.

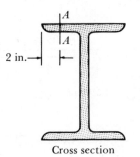

Figure 4-3.7

Cross section

4-4.1 A cast iron beam is loaded as shown in the figure. Determine the principal stresses at the three points A, B, and C caused by the applied force of 8000 lb. (Figure 4-4.1 appears at the top of page 108.)

4-4.2 A simple beam 50 mm wide by 150 mm high spans 1.5 m and supports a uniformly distributed load of 90 kN/m, including its own weight. Determine the principal stresses and their directions at points A, B, C, D, and E at the section shown in the figure (see page 108).

4-4.3 The simple beam has the cross section shown. Given that the allowable stresses are 70 MPa in shear and 110 MPa in tension and compression, what is the maximum allowable load P in newtons? (See Figure 4-4.3 on page 108.)

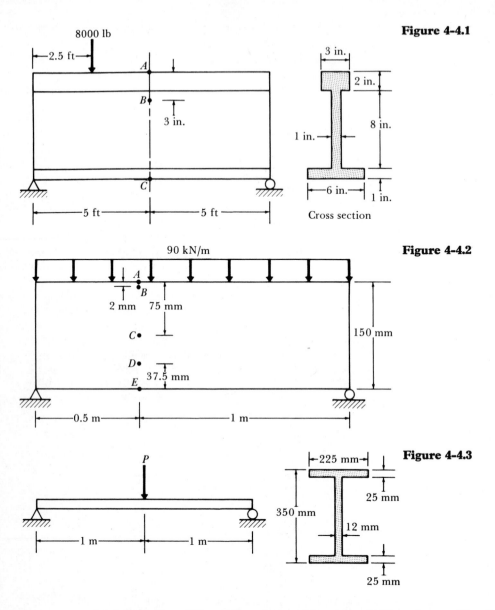

Figure 4-4.1

Figure 4-4.2

Figure 4-4.3

4-4.4 A W 24 × 130 cantilever beam with a span of 8 ft carries a uniformly distributed load of 12,000 lb/ft. Determine the principal stresses and the maximum shearing stress in the beam.

Chapter five

Lateral deflections of beams

R5-1 Lateral deflections of beams

Our task here is to derive an equation that relates the bending moment in a straight beam of constant cross section to the lateral deflections of the beam. (When we speak of the lateral deflections of a beam, we are referring to the displacement of the longitudinal centroidal axis of the beam in the direction perpendicular to the longitudinal axis of the beam.) To derive such an equation, we must go through a process of geometric and algebraic manipulation. We begin by examining a beam for which the bending moment is constant, as shown in Figure 5-1(a).

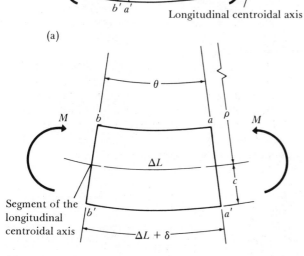

Figure 5-1

(a)

(b)

▶ **NOTE** In the case of pure bending, no shear strain exists. Consequently, planes aa' and bb' remain planes—a fact that can be experimentally verified. (Refer to limiting condition number 2 at the end of Section R4-2.)

From Figure 5-1(b), we find that

$$\Delta L = \rho\theta \tag{1a}$$

where ρ is the radius of curvature of the segment ΔL. We also obtain

$$\Delta L + \delta = (\rho + c)\theta \tag{1b}$$

Then combining Equations (1a) and (1b) gives

$$\theta = \frac{\Delta L}{\rho} = \frac{\Delta L + \delta}{\rho + c} \tag{2}$$

Now, rearranging Equation (2) and limiting our study to linearly elastic materials for which Hooke's law applies, we obtain

$$\frac{c}{\rho} = \frac{\delta}{\Delta L} = \epsilon = \frac{\sigma_c}{E} = \frac{Mc}{EI} \tag{3}$$

Don't let us slip something past you. Check each one of the preceding steps to make certain that you agree.

Now from Equation (3) we obtain the expression

$$\frac{1}{\rho} = \frac{M}{EI} \tag{4}$$

▶ **NOTE** In the more general case where shear forces exist and consequently the bending moment is not constant, we shall continue to assume that deflection due to shear is negligible compared with deflection due to bending. Thus we shall continue to assume that plane sections remain plane (see Assumption 2 in Section R4-2). Consequently, we shall assume that Equation (4) is applicable to all long slender beams. The assumption that deflections due to shear are negligible when compared to the deflections due to bending is quite good for long slender beams. As an example, the deflection due to bending of a short rectangular steel beam with length only six times its depth will be approximately 30 times the deflection due to shear. This difference increases approximately in proportion to the square of the length divided by the depth.

We are most frequently interested in lateral deflections instead of the radius of curvature. Thus it is necessary to derive a more general expression that relates the applied loads to the lateral deflections of the beam. We may obtain a more general expression by utilizing the following equation:

$$\frac{1}{\rho} = \frac{d^2y/dx^2}{[1 + (dy/dx)^2]^{3/2}} \tag{5}$$

If Equation (5) is not familiar to you, a quick check of the mathematical expression for curvature in any calculus book will refresh your memory.

In this chapter, the coordinate y is used as a measure of the lateral deflection of the centroidal axis of a beam, as shown in Figure 5-2.

Figure 5-2

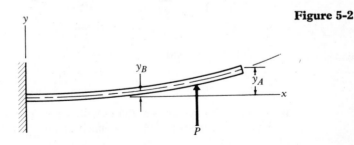

▶ **NOTE** In Chapter 4, the y coordinate was measured from the neutral axis of a beam cross section to another point on the cross section. Consequently, the meaning of coordinate y is different in Chapter 5 from what it was in Chapter 4. This particular duplication of notation is very common in texts on mechanics of materials.

When we examine the shape of the elastic curve of most beams, we shall find the slope to be very small. Thus we can make the assumption that

$$\left(\frac{dy}{dx}\right)^2 \ll 1$$

With this approximation, we can now simplify Equation (5) to the form

$$\frac{1}{\rho} \approx \frac{d^2y}{dx^2} \tag{6}$$

Hence, from Equations (4) and (6), we can obtain

$$\frac{d^2y}{dx^2} = \frac{M}{EI} \tag{7}$$

The derivation is finished, and you may breathe a sigh of relief.
■ **STOP**

You may now reestablish your normal breathing patterns and proceed with the remainder of this section.

Since the bending moment M is equal to $EI(d^2y/dx^2)$, a positive bending moment must cause the beam to bend with a positive d^2y/dx^2, as shown in Figure 5-3(a). Similarly, a beam having a negative d^2y/dx^2 and a corresponding negative bending moment is illustrated in Figure 5-3(b).

Figure 5-3

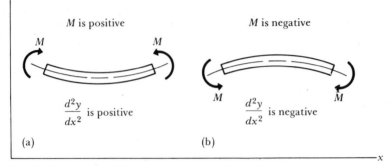

▶ **NOTE** The sign convention used here for bending moment is the same as the one we used for moment diagrams in Section R4-1.

Equation (7) is the differential equation that defines the lateral deflections of the centroidal axis of a beam with constant cross sections. This chapter is devoted to obtaining solutions to Equation (7). Don't panic if you have not had a course in differential equations, because the solution to Equation (7) is relatively simple (just requires two integrations), and we will show you how it is done.

At this point we would like to remind you of the limiting conditions on which the derivation of Equation (7) is based.

Limiting conditions

1 All the limitations related to the flexure formula as given in Section R4-2 apply, since the flexure formula was used in the derivation of Equation (7).
2 $(dy/dx)^2 \ll 1$. This implies that the slope of the longitudinal centroidal axis remains small at all points along the beam.

● **CAUTION** Remember that assumptions are more than just something necessary to derive an equation. They spell out the limitations of the theory developed. If you aren't familiar with the assumptions on which any theory is based, you run the risk of using information that is not applicable to your situation.

Do you remember from your calculus course that, when you integrated a function without limits of integration, you obtained a constant of integration?

ANSWER Yes. Great, proceed with this section.
ANSWER No. Go back to your calculus text for a review.

Now it should come as no great surprise that the solution of

$$\frac{d^2y}{dx^2} = \frac{M}{EI}$$

will involve two integrations, and consequently will contain two constants of integration. You will be able to determine these constants corresponding to a specific problem by studying the boundary conditions and/or the conditions of continuity.

Boundary conditions A boundary condition is defined as a known set of values for x and y (deflection), or x and dy/dx (slope). If a beam is pinned at $x = 0$, as shown in Figure 5-4, what is known about the deflection and slope of the beam at the pin?

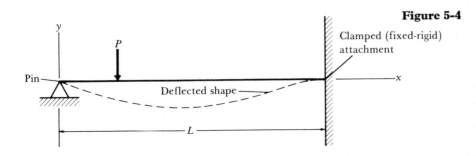

Figure 5-4

Clamped (fixed-rigid) attachment

Pin

Deflected shape

L

Your answer is correct if you say that we know the deflection y to be zero at $x = 0$, and that we have no information about the slope dy/dx at $x = 0$. Now state what you can determine about the slope and deflection at the point $x = L$ for the beam in Figure 5-4.

Your answer is correct if you said that both the slope and the deflection remain at zero at $x = L$.

In summary we have the following.

Boundary conditions

Pinned connection at $x = x_1$ (the beam is free to rotate, but not free to translate): $x = x_1, y = 0$

Clamped connection at $x = x_2$ (the beam is prevented from rotating and translating): $x = x_2, y = 0$, and $dy/dx = 0$

Continuity conditions Many beams are subjected to abrupt changes in loading along the beam, i.e., concentrated loads, reactions, or distinct changes in the amount of a uniformly distributed load. The expressions for the bending moments on the left- and right-hand sides of any abrupt change in load must be different functions of x. Thus it is impossible to write a single continuous equation for the bending moment, in terms of ordinary algebraic functions, that is valid for the entire length of the beam. This difficulty can be resolved by writing separate bending-moment equations for each segment of the beam. Figure 5-5 shows the division of a beam into segments.

Although the segments are bounded by abrupt changes in load, the deflection and slope of the longitudinal centroidal axis cannot change abruptly at any point along the length of a beam. As a consequence, we observe that at each point between adjacent segments, the deflection (and slope) of the longitudinal centroidal axis has only one value and must be the same whether the point is approached from the left or from the right. Thus we say that the continuity conditions are defined as the equality of slope and deflection at the junction of two segments, as determined by the elastic-curve equations for both segments. We can illustrate the conditions of continuity by examining the beam in Figure 5-5.

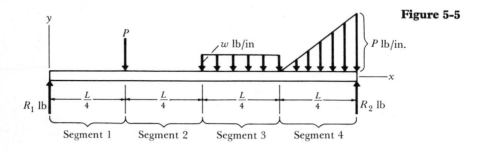

Figure 5-5

If we designate the lateral deflections in segment 1 by y_1, and the lateral deflections in segment 2 by y_2, then the conditions of continuity are

$$y_1 = y_2 \qquad \text{at} \qquad x = \frac{L}{4}$$

and

$$\frac{dy_1}{dx} = \frac{dy_2}{dx} \qquad \text{at} \qquad x = \frac{L}{4}$$

In summary we have the following continuity conditions.

Continuity conditions

$$y_1 = y_2 \quad \text{and} \quad \frac{dy_1}{dx} = \frac{dy_2}{dx} \qquad \text{at} \qquad x = x_0$$

Get up and stretch for a few minutes before turning to Section SG5-1 of the Study Guide. Then join us in solving some problems by means of Equation (7) and our newly acquired knowledge about boundary conditions and continuity conditions. ■ STOP

R5-2 Singularity functions for beam deflections

Before we get into a detailed definition of singularity functions, it may be helpful if we make a few general comments about the role of these functions as related to beam deflections. A singularity function is a mathematical tool that we shall use to describe the applied loads in a beam. We introduce the concept of singularity functions because it allows us to describe the applied loads in a beam with a single equation instead of several equations, each corresponding to a different segment. To create a function that will serve our purpose, we shall define *singularity functions*, which have properties that may initially seem meaningless to you. You may need to wait until we have used singularity functions in solving beam-deflection problems before you can fully appreciate the usefulness of these properties.

In this text, we shall always identify singularity functions by brackets: $\langle \ \rangle$. A *singularity function of x* is written in the form

$$F_n(x) = \langle x - x_0 \rangle^n \tag{8}$$

in which n is any integer (positive or negative), including zero, and x_0 is a particular value of x described in the following paragraphs.

▶ **NOTE** A variety of different notation systems are used in other texts to represent singularity functions. Regardless of the notation system, the mathematical meaning is the same.

For our purposes, singularity functions have the following properties:

$$n \geq 0: \qquad \langle x - x_0 \rangle^n = \begin{cases} (x - x_0)^n & \text{when } x \geq x_0 \\ 0 & \text{when } x < x_0 \end{cases} \tag{9a}$$

$$n \geq 0: \qquad \int \langle x - x_0 \rangle^n \, dx = \frac{1}{n + 1} \langle x - x_0 \rangle^{n+1} \tag{9b}$$

$$n > 0: \qquad \frac{d}{dx} \langle x - x_0 \rangle^n = n \langle x - x_0 \rangle^{n-1} \tag{9c}$$

$$n < 0: \qquad \langle x - x_0 \rangle^n = \begin{cases} \infty & \text{when } x = x_0 \\ 0 & \text{when } x \neq x_0 \end{cases} \tag{10a}$$

$$n \leq 0: \qquad \int \langle x - x_0 \rangle^n \, dx = \langle x - x_0 \rangle^{n+1} \tag{10b}$$

$$n \leq 0: \qquad \frac{d}{dx} \langle x - x_0 \rangle^n = \langle x - x_0 \rangle^{n-1} \tag{10c}$$

● **CAUTION** Be sure to observe that, when n is negative or zero, integration and differentiation result in no change in the coefficients of the singularity functions.

Let's consider the application of singularity functions to several common types of load.

1 *Ramp load* (Figure 5-6). The equation for line *aa* is

$$w = \frac{p}{L - x_0} (x - x_0)$$

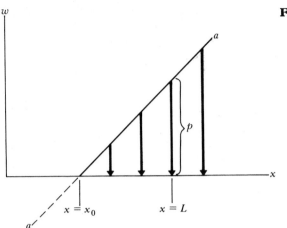

Figure 5-6

The equation for line aa is not precisely the same as the equation for the ramp load, since line aa exists for $x < x_0$ and the ramp load does not. According to the previously described properties of singularity functions, the equation for a ramp load is

$$w = \frac{p}{L - x_0} \langle x - x_0 \rangle^1 \tag{11}$$

This equation is valid because, from Equation (9a), we have

$$\frac{p}{L - x_0} \langle x - x_0 \rangle^1 = \begin{cases} \dfrac{p}{L - x_0} (x - x_0) & \text{for } x \geq x_0 \\ 0 & \text{for } x < x_0 \end{cases}$$

2 *Uniformly distributed load* (Figure 5-7). The uniformly distributed load in Figure 5-7 can be described by the equation

$$w = p \langle x - x_0 \rangle^0 \tag{12}$$

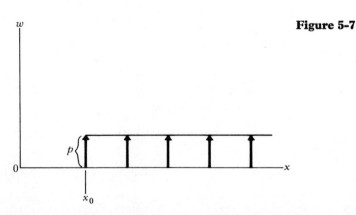

Figure 5-7

Since, from Equation (9a), we have

$$p\langle x - x_0\rangle^0 = \begin{cases} p(x - x_0)^0 = p & \text{for } x \geq x_0 \\ 0 & \text{for } x < x_0 \end{cases}$$

3 *Concentrated load* (point load). In reality it is impossible for a load to act at a single point. All real loads must act over at least some small area. We represent a concentrated load P by a distributed load of large intensity acting over a very small dimension Δ, as shown in Figure 5-8.

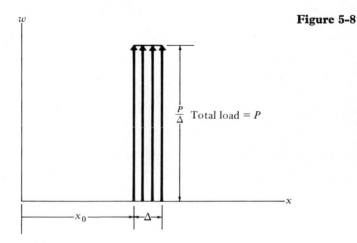

Figure 5-8

Observe that the total load in Figure 5-8 is $(P/\Delta)(\Delta) = P$, which indeed is the magnitude of the concentrated load it is to represent. In the limit, as Δ approaches zero, the distributed load intensity P/Δ will approach infinity, and we will have created the mathematical representation of a concentrated load:

$$w = \lim_{\Delta \to 0} \frac{P}{\Delta}$$

As $\Delta \to 0$, we can replace Δ by $(x - x_0)$:

$$w = \lim_{\Delta \to 0} \frac{P}{\Delta} = \lim_{x \to x_0} \frac{P}{(x - x_0)}$$

Thus the concentrated load, shown in Figure 5-9, is described by the singularity function

$$w = P\langle x - x_0\rangle^{-1} \tag{13}$$

We observe from Equation (10a) that the load intensity of

$$w = P\langle x - x_0\rangle^{-1}$$

is infinitely large at $x = x_0$. However, with a concentrated load, we are really interested only in the total applied load, not the fictitious load intensity. And the net result of an infinitely large load intensity acting over an infinitesimally small length yielding a finite concentrated load is our primary concern.

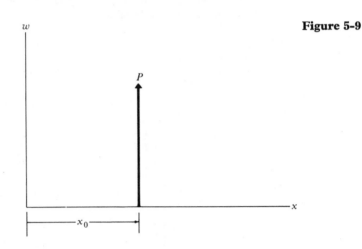

Figure 5-9

4 *Concentrated couple.* To determine the singularity function that describes a concentrated couple of magnitude c, we begin with two oppositely directed distributed loads of large intensity, each acting over a very small dimension Δ, as shown in Figure 5-10. We observe from Figure 5-10 that the magnitude of the couple created by the two oppositely directed distributed loads is

$$\sum M_{x_0} = 2 \left[(w)(\Delta) \left(\frac{\Delta}{2} \right) \right] = c$$

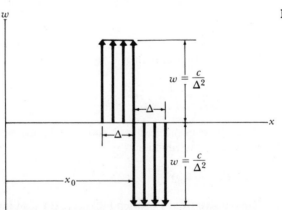

Figure 5-10

If we let

$$w = \frac{c}{\Delta^2} \tag{14}$$

then we have

$$\sum M_{x_0} = 2\left[\left(\frac{c}{\Delta^2}\right)(\Delta)\left(\frac{\Delta}{2}\right)\right] = c$$

which indeed is the magnitude of the couple we wish to represent. As $\Delta \to 0$, we can replace Δ by $\langle x - x_0 \rangle^1$, and Equation (14) becomes

$$w = \lim_{\Delta \to 0} \frac{c}{\Delta^2} = \lim_{x \to x_0} \frac{c}{(x - x_0)^2}$$

Thus the concentrated moment, shown in Figure 5-11, is described by the singularity function

$$w = c\langle x - x_0 \rangle^{-2} \tag{15}$$

Figure 5-11

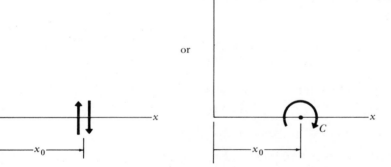

or

Figure 5-12 shows the most common kinds of load and their corresponding singularity functions.

Sign conventions

The sign convention for applied loads is: Upward forces and clockwise couples bear the plus sign, whereas downward forces and counterclockwise couples bear the minus sign.

Clockwise couple of magnitude c applied at $x = x_0$.

Figure 5-12(a)

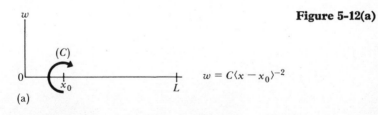

$$w = C\langle x - x_0 \rangle^{-2}$$

(a)

Concentrated load P applied upward at $x = x_0$.   **Figure 5-12(b)**

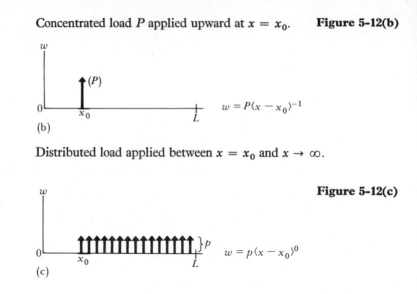

$$w = P\langle x - x_0 \rangle^{-1}$$

(b)

Distributed load applied between $x = x_0$ and $x \rightarrow \infty$.

Figure 5-12(c)

$$w = p\langle x - x_0 \rangle^0$$

(c)

Distributed load of magnitude p applied in the interval $x_0 \leq x \leq x_1$.

Figure 5-12(d)

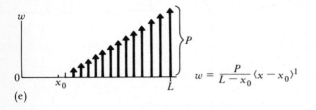

$$w = p\langle x - x_0 \rangle^0 - p\langle x - x_1 \rangle^0$$

(d)

Ramp load of slope $p/(L - x_0)$ applied between $x = x_0$ and $x \rightarrow \infty$.

Figure 5-12(e)

$$w = \frac{P}{L - x_0}\langle x - x_0 \rangle^1$$

(e)

Ramp load of slope $p/(L - x_0)$ applied in the interval $x_0 \leq x \leq x_1$.

Figure 5-12(f)

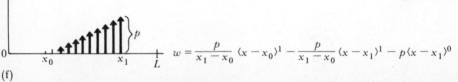

$$w = \frac{p}{x_1 - x_0}\langle x - x_0 \rangle^1 - \frac{p}{x_1 - x_0}\langle x - x_1 \rangle^1 - p\langle x - x_1 \rangle^0$$

(f)

▶ **NOTE** The coefficient of each singularity function is a constant. Every such coefficient is considered to be the "strength" of the particular singularity. Thus, as an example, the couple c has units of force times length and c is the magnitude of the couple [Figures 5-11 and 5-12(a)]; P is the magnitude of the concentrated load and has units of force [Figures 5-9 and 5-12(b)]; and the intensity of the step is p with units of force per unit length [Figures 5-7 and 5-12(c)].

Two of the preceding examples need special mention. The sketch in Figure 5-13 shows how we obtained the singularity function for a uniform load existing over a portion of the beam, as given in Figure 5-12(d). Note that the process illustrated in Figure 5-13 simply involves the superposition of the uniform load $-p$ beginning at x_1 on the uniform load p beginning at x_0.

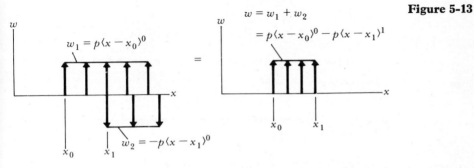

Figure 5-13

The singularity function for the ramp load illustrated in Figure 5-12(f) is derived by the process of superposition shown in Figure 5-14. Let's assume that the magnitude of the ramp load is p (lb/in.) at the point x_1.

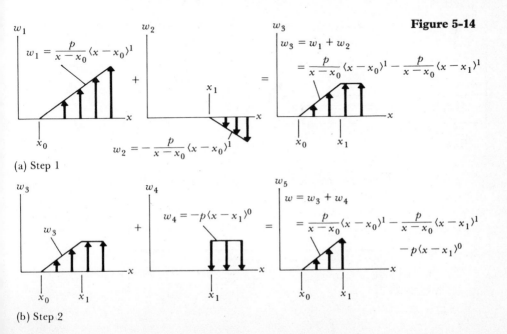

Figure 5-14

(a) Step 1

(b) Step 2

The following is a simple example to help tie this section together.

EXAMPLE Write a singularity function that describes the loads applied to the beam shown in Figure 5-15. The length of the beam is ℓ (in.).

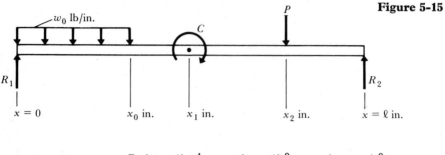

Figure 5-15

$$w = R_1 \langle x - 0 \rangle^{-1} - w_0 \langle x - 0 \rangle^0 + w_0 \langle x - x_0 \rangle^0 \\ + c \langle x - x_1 \rangle^{-2} - P \langle x - x_2 \rangle^{-1} + R_2 \langle x - \ell \rangle^{-1}$$

Now turn to Section SG5-2 of the Study Guide. ■ **STOP**

Problems

Assume EI = constant unless otherwise noted.

5-1.1 Determine the deflection and the slope of the free end of the given cantilever beam.

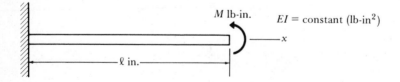

Figure 5-1.1

5-1.2 Determine the slope and the deflection of the following beam at point C.

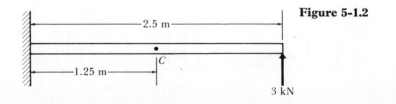

Figure 5-1.2

5-1.3 Locate the point of maximum deflection between the supports.

Figure 5-1.3

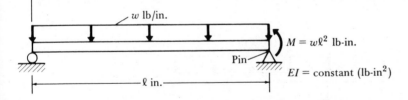

w lb/in.

$M = w\ell^2$ lb-in.

Pin

$EI = $ constant (lb-in^2)

ℓ in.

5-1.4 Determine the maximum deflection of the following beam.

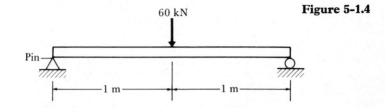

60 kN

Figure 5-1.4

Pin

1 m · 1 m

5-1.5 The given cantilever beam supports a triangularly distributed load of maximum intensity P_0. Determine the deflection and the angle of rotation (slope) of the free end.

Figure 5-1.5

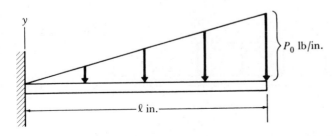

P_0 lb/in.

ℓ in.

5-1.6 Determine the expression for the deflections of the following beam.

Figure 5-1.6

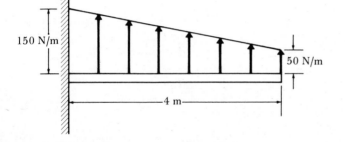

150 N/m

50 N/m

4 m

5-1.7 Determine the deflections of the given W 8 × 20 cantilever beam at points A and B. The beam is made of structural steel.

Figure 5-1.7

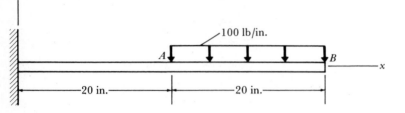

5-1.8 Determine the slope of the following beam at the point where the couple $c = 10$ kN · m is applied.

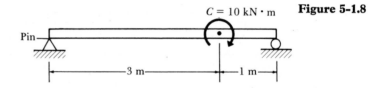

Figure 5-1.8

5-1.9 Determine the deflection of the right-hand end of the following cantilever beam by the method of superposition.

Figure 5-1.9

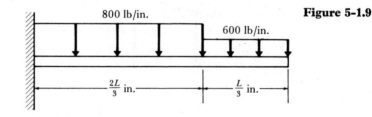

5-1.10 Determine the deflection at midspan ($x = L/2$) of the following cantilever beam by the method of superposition.

Figure 5-1.10

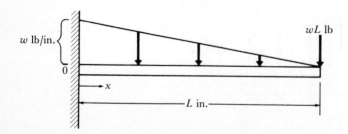

5-1.11 Compute the deflection and the slope of the following beam at point B.

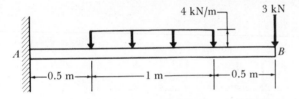

Figure 5-1.11

5-1.12 Use the method of superposition to determine the deflection at the midspan of the following beam.

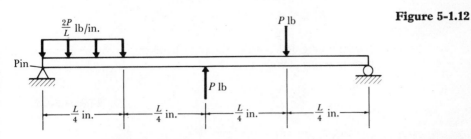

Figure 5-1.12

5-1.13 Use the method of superposition to determine the maximum deflection of the following beam.

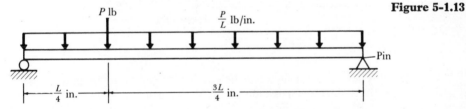

Figure 5-1.13

5-1.14 Determine the slope of the following beam at point B.

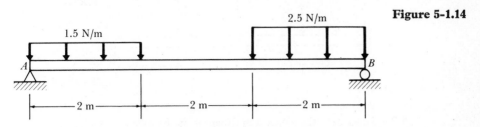

Figure 5-1.14

5-1.15 Determine the deflection of the following beam at point C.

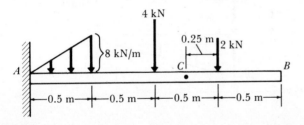

Figure 5-1.15

5-2.1 Determine the deflection function for the following beam by using the singularity functions.

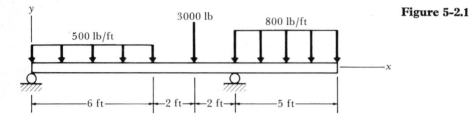

Figure 5-2.1

5-2.2 Determine the deflection of the right-hand end of the given cantilever beam in terms of EI.

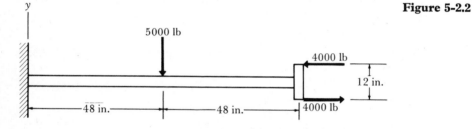

Figure 5-2.2

5-2.3 The maximum allowable deflection of the following beam is $10^4/EI$. Determine the maximum allowable value of P.

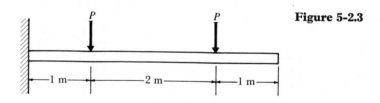

Figure 5-2.3

5-2.4 Determine the slope at midspan of the following beam.

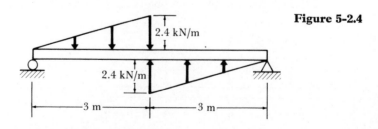

Figure 5-2.4

5-2.5 Calculate the deflection at midspan of the following beam.

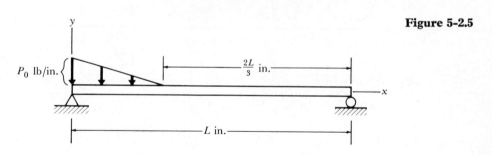

Figure 5-2.5

5-2.6 Determine the deflection of the following beam at point B.

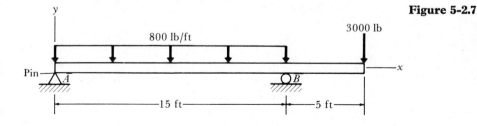

Figure 5-2.6

5-2.7 Use singularity functions to determine the maximum deflection of the following beam between A and B.

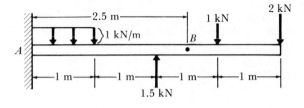

Figure 5-2.7

Chapter six

Combined loading

R6-1 Combined loading

Stresses resulting from axial, torsional, and flexural loads were investigated in Chapters 1, 3, and 4, respectively. Let's review the expression used for determining stresses in each of these cases, as well as the corresponding limiting conditions.

Axial loading (Chapter 1)

$$\sigma = \frac{P}{A} \tag{1}$$

where:

σ = uniform normal stress
A = area on which the stress σ acts
P = resultant force resisted by the area A and perpendicular to A

For the normal stress to be of uniform intensity over the resisting area A [that is, for Equation (1) to be valid], the following conditions must exist.

1 The member on which the load acts must be straight, with a constant cross section.
2 The load must be applied through the centroid of the cross section.
3 The member must be made from a homogeneous material.

Torsional loading (Chapter 3)

$$\tau = \frac{T\rho}{J} \tag{2}$$

where:

τ = torsional shear stress
J = polar moment of inertia of the cross-sectional area
T = internal torque resisted by the cross-sectional area where the stress τ is to be determined

ρ = the radial distance measured from the center of the cross section to the point at which τ is to be determined

The following assumptions and corresponding limiting conditions were involved in the derivation of Equation (2) for torsional stress.

1 Plane sections before twisting must remain plane after twisting, and radial lines must remain straight—a condition that holds only for circular sections.
2 All longitudinal fibers must have the same length—a condition that holds true only for straight members of constant diameter.
3 The modulus of rigidity G must have the same value at all points and in all directions—the material must be homogeneous, isotropic, and linearly elastic.

Flexural loading (Chapter 4)

$$\sigma = \frac{My}{I} \tag{3}$$

where:

σ = normal stress at a point in the axial direction of the beam
M = the internal bending moment acting about the centroidal axis of the cross section at which the stress is to be determined
y = the perpendicular distance measured from the centroidal axis of the cross section to the point at which the stress is to be determined
I = the moment of inertia of the cross-sectional area about the centroidal axis

The assumptions and corresponding limiting conditions of the flexure formula given by Equation (3) are as follows.

1 The beam cross section must have an axis of symmetry in the plane of the loads.
2 Cross sections must remain plane after bending, which means that the cross sections must be free of shear forces (this requirement is not usually significant unless the beam is short and deep).
3 All longitudinal fibers must have the same length, which means that the beam must be straight with a constant cross section.
4 Hooke's law must apply (the stress must not exceed the proportional limit of the material.)
5 The modulus of elasticity must be the same for all longitudinal fibers and must be the same in tension and compression.

$$\tau = \frac{VQ}{It} \tag{4}$$

where:

τ = shear stress (horizontal shear stress = vertical shear stress)

V = internal shear force acting on the cross section where τ is to be determined

$Q = \int_{A_F} y\, dA$, the statical or first moment of the area A_F about the centroidal axis

I = moment of inertia of the cross-sectional area about the centroidal axis [the same as in Equation (3)]

t = the width of the plane on which τ acts

The assumptions and corresponding limiting conditions of Equation (4) are as follows.

1–5 The same as in Equation (3).
  6 The stress is assumed to be uniform over the thickness t. Thus t should be small compared with the other dimensions of the cross section, and the sides of the beam should be parallel.

Up to this point in our study of mechanics of materials, we have investigated structural members subjected to axial loads or torsional loads or flexural loads. Many structural elements and machine parts, however, are subjected to a combination of any two or all three of these types of loads. Stresses resulting from combined loading can be obtained by the principle of superposition, *provided that the combined stresses do not exceed the proportional limit.*

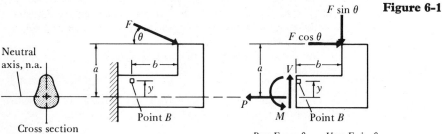

Figure 6-1

$$P = F \cos \theta \qquad V = F \sin \theta$$
$$M = Fa \cos \theta + Fb \sin \theta$$

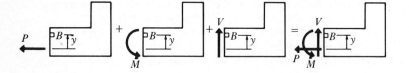

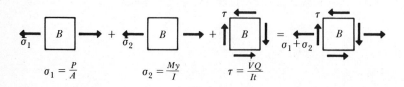

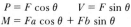

$$\sigma_1 = \frac{P}{A} \qquad \sigma_2 = \frac{My}{I} \qquad \tau = \frac{VQ}{It}$$

As an example, the cross-sectional area at point B in Figure 6-1 resists an axial load P, a bending moment M, and a shear force V. Thus it is subjected to a combination of axial loading and flexural loading. The sequence of free-body diagrams in Figure 6-1 shows how we can use our knowledge of the ways to find stresses from axial loads and from flexural loads, along with the principle of superposition, to determine the combined stresses at point B.

The cross-sectional area at points B and D in Figure 6-1 (D is a point on the side of the member, and B is on the top) resists an axial load, a shear force, a bending moment, and a torque. Thus it is subjected to a combination of axial loading, flexural loading, and torsional loading. Figure 6-2 on page 132 shows how we can use our knowledge of the ways to find stresses corresponding to the individual load types, along with the principle of superposition, to obtain the combined stresses at points B and D.

After carefully studying Figures 6-1 and 6-2, turn to the Study Guide where you will find more explanation and examples. ■ **STOP**

Problems

6-1.1 A post is loaded by two forces, as shown. Determine the normal stresses at corners A, B, C, and D acting on cross section $ABCD$. Also determine the locus of points of zero normal stresses on cross section $ABCD$.

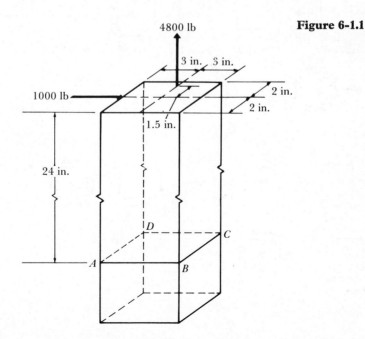

Figure 6-1.1

Figure 6-2

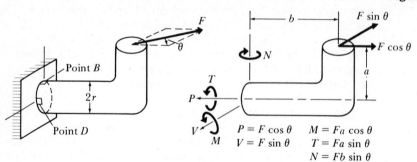

$$P = F \cos \theta \qquad M = Fa \cos \theta$$
$$V = F \sin \theta \qquad T = Fa \sin \theta$$
$$N = Fb \sin \theta$$

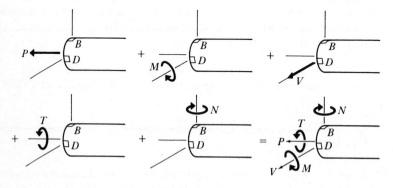

For point C

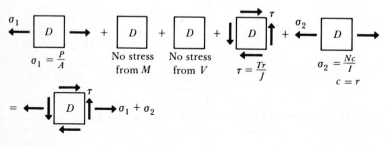

$$\sigma_1 = \frac{P}{A} \qquad \text{No stress} \qquad \text{No stress} \qquad \tau = \frac{Tr}{J} \qquad \sigma_2 = \frac{Nc}{I}$$
$$\text{from } M \qquad \text{from } V \qquad \qquad \qquad c = r$$

$$= \rightarrow \sigma_1 + \sigma_2$$

For point B

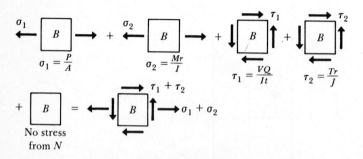

$$\sigma_1 = \frac{P}{A} \qquad \sigma_2 = \frac{Mr}{I} \qquad \tau_1 = \frac{VQ}{It} \qquad \tau_2 = \frac{Tr}{J}$$

$$+ \quad B \quad = \quad \rightarrow \sigma_1 + \sigma_2$$

No stress
from N

6-1.2 A press has a cast-iron frame in the shape as shown. The allowable stresses on cross section $A-A$ are 34.5-MPa tension and 124-MPa compression. Determine the capacity P of the press.

Figure 6-1.2

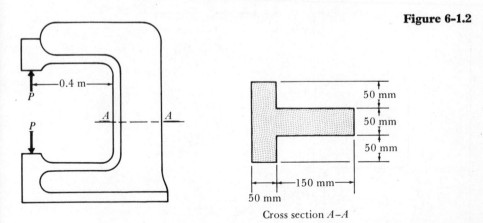

Cross section $A-A$

6-1.3 Determine the principal stresses at point A on the bracket, and show them on a properly oriented element. Also determine the maximum shear stress at point A. The cross section is a rectangular 50 mm × 150 mm section.

Figure 6-1.3

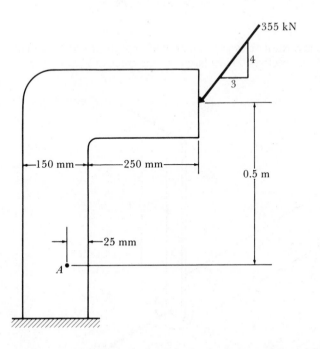

6-1.4 The hoist is supporting a 20,000-lb load at the end of a cable. Determine the state of stress at point A.

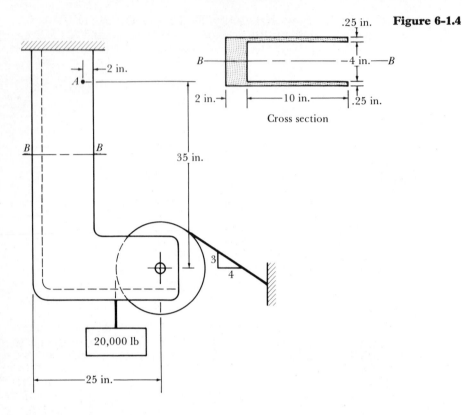

.25 in.　**Figure 6-1.4**

B

−4 in.—B

−2 in.

A

2 in.→ |—10 in.—| .25 in.

Cross section

B | B

35 in.

3
4

20,000 lb

25 in.

6-2.1 The tensile and shear stresses at point A are not to exceed 138 MPa and 55 MPa, respectively. How large can the value of P be on the shaft shown?

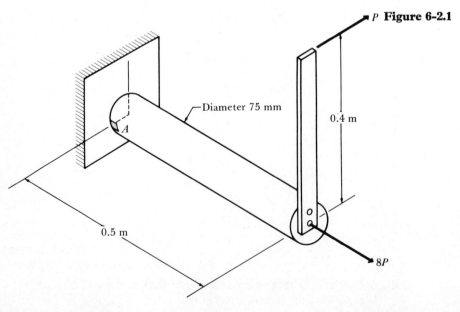

P **Figure 6-2.1**

—Diameter 75 mm

0.4 m

0.5 m

8P

6-2.2 The bracket is loaded with a 500-lb force, as shown (the force is in the xy plane, and inclined at 35° from the x axis). Determine the principal stresses and the maximum shear stress at point A, and show them on a properly oriented element(s). Point A is on the top of the bracket.

Figure 6-2.2

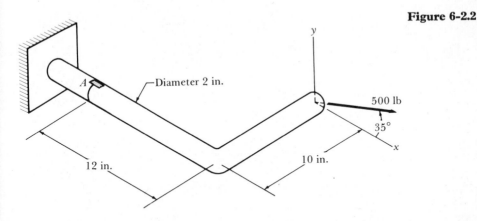

6-2.3 A 2-in.-diameter shaft supports gears at A and B, each gear having a 10-in. radius. The gear at A is driven by another gear (not shown) which exerts on it a vertical tangential force (upward) of 600 lb. The gear at B drives another gear (not shown) which exerts on it a vertical tangential force (downward) of 600 lb. Determine the principal stresses and maximum shear stress in the shaft. Assume the bearings at D and C to be simple supports.

Figure 6-2.3

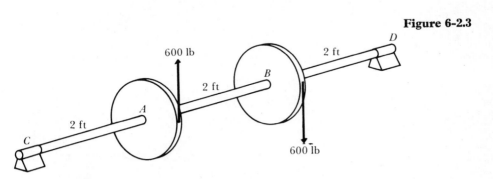

6-2.4 A steel pressure vessel 250 mm in diameter, with a 6.4-mm wall, also acts as an eccentrically loaded simply supported beam. The internal pressure is 2.8 MPa, and the mass W is 13.3×10^3 kg. Determine the state of stress at point A, and show your answer on an infinitesimal element. Determine the principal stresses and maximum shear stress. (Neglect the mass of the vessel.)

Figure 6-2.4

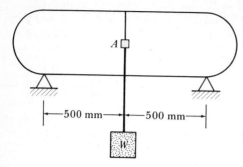

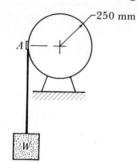

Chapter seven

Statically indeterminate structures

R7-1 Statically indeterminate structures

Degree of static indeterminacy

If n_R is the number of unknown reactions and/or deflections, and n_E is the number of possible independent equations corresponding to a particular structural element, then we say that the structural element is *statically indeterminate to the degree n*, where $n = n_R - n_E$. Stated in another way:

If we are to determine the reactions for a statically indeterminate structure, we need to obtain n independent equations in addition to the equations of equilibrium.

Thus, if we are to know how many additional independent equations we must derive by the methods of mechanics of materials, it is important that we know how many independent equations of equilibrium can be written for each type of structural system.

State the number of independent equations of equilibrium that can be written corresponding to rigid bodies under the action of each of the following loading conditions.

1 Collinear-force system
2 Coplanar-concurrent-force system
3 Three-dimensional concurrent-force system
4 Coplanar-parallel-force system
5 Three-dimensional parallel-force system
6 Coplanar nonconcurrent system
7 Three-dimensional nonconcurrent-force system

The correct answers are given in Table 7-1.

If your answers match those given in Table 7-1 and you feel confident that you can write the independent equations of equilibrium for any rigid body, then turn to Section SG7-1 of the Study Guide. If all of your answers are not correct and/or you feel the need for a quick review of the types of equilibrium equations that can be written to correspond to rigid bodies, then study the following material.

| | Number of Table 7-1 |
Force system	independent equations of equilibrium
Collinear	1
Coplanar concurrent	2
Three-dimensional concurrent	3
Coplanar parallel	2
Three-dimensional parallel	3
Coplanar nonconcurrent	3
Three-dimensional nonconcurrent	6

Collinear-force systems

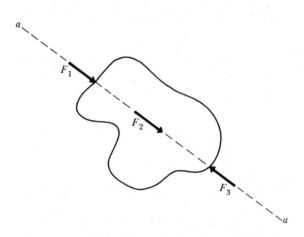

In the case of a collinear-force system, the only independent equation of equilibrium is

$$\sum F_{aa} = 0 \qquad \text{(One independent equation)}$$

Coplanar-concurrent-force systems

▶ **NOTE** The xy coordinate axes and the loads F_1, F_2, F_3, F_4, F_5 are all contained in the same plane.

The independent equations of equilibrium corresponding to a coplanar-concurrent-force system are

$$\sum F_x = 0$$

$$\sum F_y = 0 \qquad \text{(Two independent equations)}$$

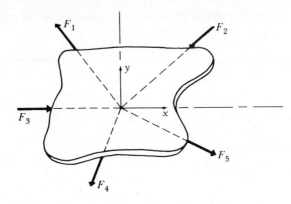

The xy coordinate system can be any rectangular coordinate system contained in the plane of the applied loads.

Three-dimensional concurrent-force systems

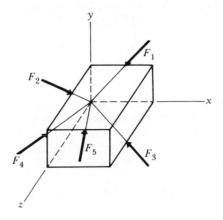

In the case of a three-dimensional concurrent-force system, the independent equations of equilibrium are

$$\sum F_x = 0$$

$$\sum F_y = 0 \qquad \text{(Three independent equations)}$$

$$\sum F_z = 0$$

The xyz coordinate system can be any three-dimensional orthogonal coordinate system.

Coplanar-parallel-force systems

For a coplanar-parallel-force system, the independent equations of equilibrium can be written in the following two ways:

1. $\sum F_y = 0$

   (Two independent equations)

   $\sum M_A = 0$

where the y axis is parallel to the applied forces and A is any point in the plane that contains the applied loads.

2. $\sum M_A = 0$

   (Two independent equations)

   $\sum M_B = 0$

where A and B are points in the plane that contains the applied loads, and the line connecting A and B is not parallel to the applied forces.

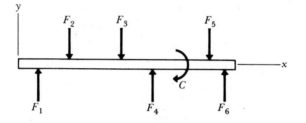

● **CAUTION** A maximum of two independent scalar equations of equilibrium can be written to correspond to the preceding problem. Do not combine the two sets of equations (1) and (2) in an attempt to obtain more than two independent equations. Any additional equations you obtain will be dependent equations.

Three-dimensional parallel-force systems

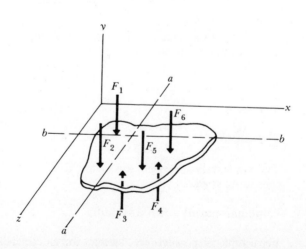

For this type of problem, we can obtain the equations of equilibrium in the following manner:

$$\sum F_y = 0$$

$$\sum M_{aa} = 0 \qquad \text{(Three independent equations)}$$

$$\sum M_{bb} = 0$$

where y is the coordinate axis parallel to the applied forces, lines aa and bb are in or parallel to the xz plane, and lines aa and bb are not parallel to each other.

Coplanar-nonconcurrent-force systems

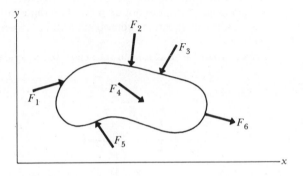

The coplanar-nonconcurrent-force system is the most general class of two-dimensional force systems. The equations of equilibrium can be obtained in the following three ways.

1. $\sum M_A = 0$

$$\sum F_x = 0 \qquad \text{(Three independent equations)}$$

$$\sum F_y = 0$$

where the rectangular coordinate system xy and the point A are contained in the plane that also contains the applied loads.

Sometimes it is convenient to replace either or both of the preceding force equations by moment equations in the following two ways.

2. $\sum M_A = 0$

$$\sum M_B = 0 \qquad \text{(Three independent equations)}$$

$$\sum F_x = 0$$

where the moment centers A and B and the applied forces are contained in the same plane, and the line connecting A with B is not perpendicular to the x axis.

3. $\sum M_A = 0$

$\sum M_B = 0$ (Three independent equations)

$\sum M_C = 0$

where the moment centers A, B, and C, and the applied forces are contained in the same plane, and points A, B, and C are not collinear.

● **CAUTION** A maximum of three independent scalar equations of equilibrium can be written to correspond to a nonconcurrent coplanar-force system, because only three equations are required to ensure equilibrium. Do not combine the three sets of equations (1), (2), and (3) in an attempt to obtain more than three independent equations. Any additional equations you obtain will be dependent equations.

Three-dimensional nonconcurrent-force systems

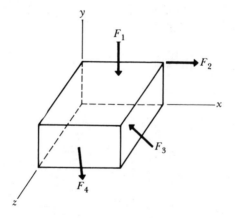

The independent equations of equilibrium are:

$\sum F_x = 0$

$\sum F_y = 0$

$\sum F_z = 0$

                         (Six independent equations)

$\sum M_x = 0$

$\sum M_y = 0$

$\sum M_z = 0$

where x, y, z are the axes corresponding to any rectangular coordinate system. It is possible to generate variant forms of these six equations in the same manner that we developed various forms of the equilibrium equations for coplanar-noncurrent-force systems. However, the six listed equations should be sufficient for our use in this text. If you have the information presented in this section so far firmly fixed in your mind, then turn to Section SG7-1 of the Study Guide. If not, then study the material again. ■ STOP

Temperature effects

Most engineering materials that are not constrained expand when heated and contract when cooled. We designate the strain due to a 1° temperature change by α, and call it the *coefficient of thermal expansion*. Consequently we can write the strain due to a temperature change of ΔT degrees as $\epsilon = \alpha \, \Delta T$.

The coefficient of thermal expansion remains approximately constant over a wide range of temperatures (in general, the coefficient of thermal expansion increases very slightly with an increase of temperature). In the case of homogeneous isotropic material, the same coefficient of thermal expansion applies at all points and in all directions. Illustrative values of the coefficient of thermal expansion for several materials are included in Appendix 2.

When temperature changes take place in a member that is restrained from free movement, thermal stresses are induced. Turn to Example 3 in Section SG7-1 of the Study Guide to learn how such thermal stresses may be determined. ■ STOP

Problems

7-1.1 A square reinforced-concrete pier 1 ft × 1 ft in cross section and 5 ft high is loaded as shown. The concrete is reinforced by 1 in. × 1 in. square steel reinforcing bars placed symmetrically about the vertical axis of the pier. Determine the stress and the deflection in both the steel and the concrete. Assume that $E_c = 2.5 \times 10^6$ psi and $E_s = 30 \times 10^6$ psi.

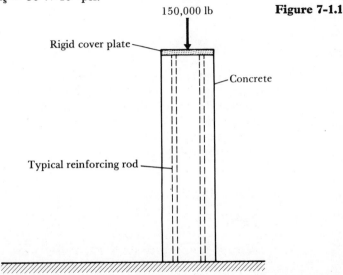

150,000 lb

Figure 7-1.1

Rigid cover plate

Concrete

Typical reinforcing rod

7-1.2 Determine the reactions at A and B in the following structural member.

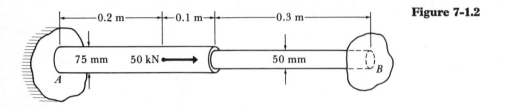

Figure 7-1.2

7-1.3 The following figure shows a pendulum that has a 10-lb weight suspended by three rods 25 in. long. Two of the rods are made of aluminum and the other of steel. Determine the load carried by each rod.

Figure 7-1.3

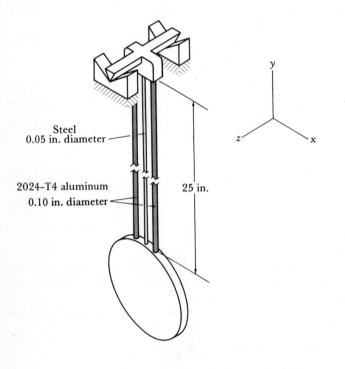

7-1.4 The assembly in the figure consists of a steel bar, a rigid block, and a brass bar securely fastened together and fastened to rigid supports at the ends. Initially there is no stress in the members. The temperature drops 30° F, and a load P of 35 kip is applied. Determine the maximum normal stresses in A and B under these conditions. Use the following data.

Bar	Cross-sectional area, in.^{2}	Length, in.	Young's modulus, ksi	Coefficient of expansion, per °F
Steel	1.0	100	30×10^3	6.5×10^{-6}
Brass	3.0	15	15×10^3	10.0×10^{-6}

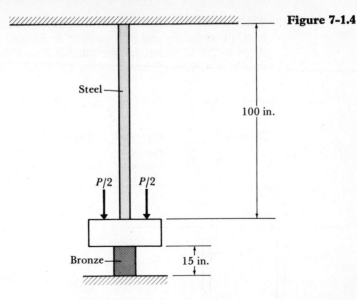

Figure 7-1.4

7-1.5 A brass rod with $d = 50$ mm is placed inside a hollow steel pipe with $d_o = 75$ mm and $d_i = 65$ mm, when $T = 20°$ C. Determine the axial loads in the brass and steel when $T = 50°$ C. Both the brass and steel members are attached rigidly to the two end plates.

	Modulus of rigidity, GPa	Young's modulus, GPa	Coefficient of expansion, per °C
Brass	42	104	18×10^{-6}
Steel	83	207	11.7×10^{-6}

Figure 7-1.5

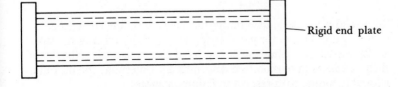

Rigid end plate

7-1.6 The member ABC is supported by steel cables ($E = 207$ GPa) at points B and C. Determine the loads in the cables at B and C. Assume member ABC to be rigid (i.e., you can neglect any deflection of ABC due to bending and consider only the rotation about A.)

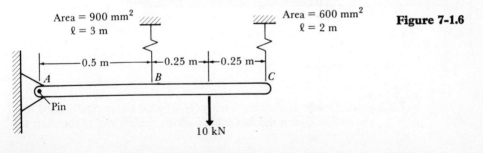

Area = 900 mm² ℓ = 3 m   Area = 600 mm² ℓ = 2 m   Figure 7-1.6

0.5 m — 0.25 m — 0.25 m

A   B   C

Pin

10 kN

7-1.7 The system in the figure is initially unstressed. Bar C is rigid and is 0.002 in. above the cast iron member D. Assume that the temperature drops by 20° F. Determine the value of P such that the stresses in members A, B, and D have the same magnitude. Use the following data.

Member	Material	Area, in^2	E, psi	Coefficient of expansion, per °F
A	Aluminum alloy	1	10×10^6	12.5×10^{-6}
B	Aluminum alloy	1	10×10^6	12.5×10^{-6}
D	Cast iron	3	14×10^6	6.6×10^{-6}

Figure 7-1.7

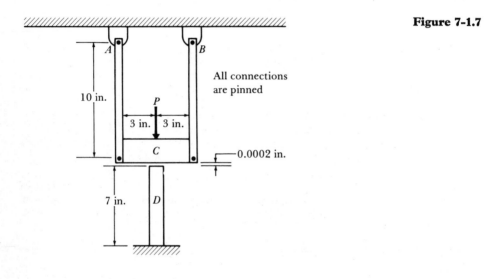

10 in.

P

3 in. | 3 in.

A B

All connections are pinned

C

0.0002 in.

7 in. D

7-2.1 Two solid circular shafts 3 in. in diameter are rigidly connected, as shown. An unknown torque T is applied at the junction of the two shafts. Assume the shearing strengths of these materials to be 0.6 of their strengths in tension (see Appendix 2). Determine the maximum allowable torque with respect to failure in shear.

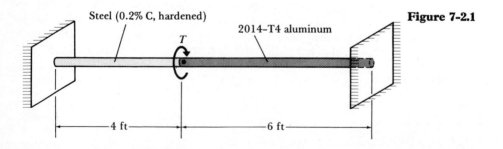

Steel (0.2% C, hardened)

2014–T4 aluminum

Figure 7-2.1

T

4 ft ————— 6 ft

7-2.2 A shaft made of aluminum alloy ($G = 4000$ ksi) is rigidly fixed to the wall at C, but the flange at A allows the left end of the shaft to rotate 0.016 rad before the rigid support of the bolts comes into play. Determine the maximum torque that can be applied at section B, given that the shearing stress is not to exceed 6 ksi in either section of the shaft.

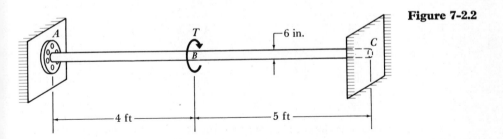

Figure 7-2.2

7-2.3 A steel shaft 7 ft long extends through and is attached to a hollow bronze shaft 4 ft long. Both shafts are fixed at the wall. Using 6000 ksi for the modulus of rigidity of bronze and 11,600 ksi for steel, determine (a) the maximum shearing stress in each material, (b) the angle of twist of the right-hand end.

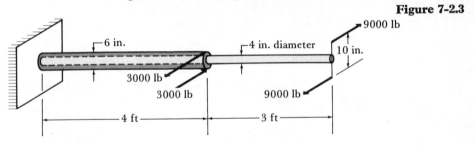

Figure 7-2.3

7-2.4 Determine the maximum allowable torque T, given that the allowable shearing stresses are 5 ksi for brass and 12 ksi for steel.

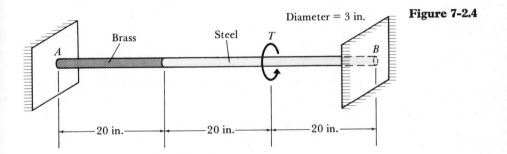

Figure 7-2.4

7-2.5 Two torsional loads of equal magnitude, 860 N·m, and opposite sense are applied to opposite ends of the structural component given in Problem 7-1.5. Determine the maximum shear stress in the brass rod.

7-2.6 Determine the reactions at A and B of the following circular rod, whose diameter is 75 mm.

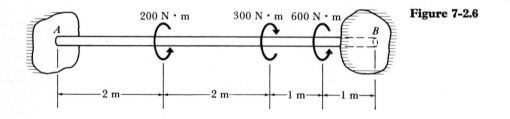

Figure 7-2.6

7-2.7 Determine the angle of twist at point C of the following circular rod, whose diameter is 50 mm.

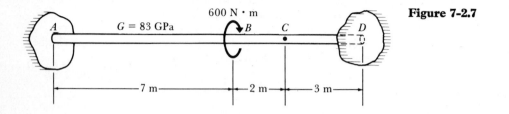

Figure 7-2.7

7-3.1 Determine the reactions of the following beam at points A and B in terms of the constant EI.

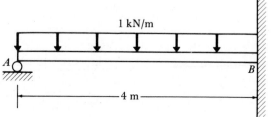

Figure 7-3.1

7-3.2 Determine the reactions of the following beam at points A, B, and C.

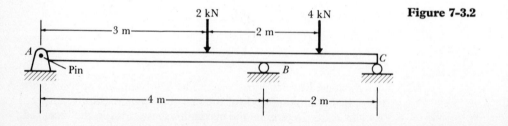

Figure 7-3.2

7-3.3 Determine the force Q (in terms of w and L) that will cause the beam to be horizontal at the right-hand end.

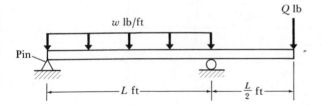

Figure 7-3.3

7-3.4 A bridge is supported on rollers at both ends and is supported by a pontoon in the center. It is loaded by a uniformly distributed load w (in pounds per unit length). How much will the pontoon sink into the water?

Let the length of the bridge be L, the weight per unit volume of water be γ, and the water-line area of the pontoon be A.

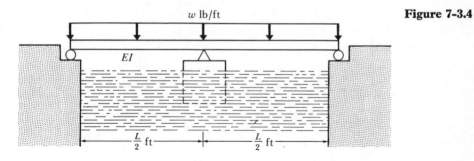

Figure 7-3.4

7-3.5 Determine the reactions at A and B of the following beam in terms of P, E, and I.

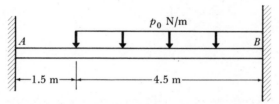

Figure 7-3.5

7-3.6 Beams A and B are made of steel. The moments of inertia of B and

A are equal. The reaction at *C* is zero before the load *w* is applied. Determine the reaction at *C* of beam *B* in terms of *w* and *L*.

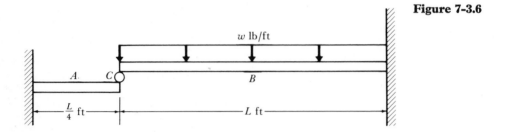

Figure 7-3.6

7-4.1 Before the load *P* was applied to the beam, the beam had been straight and the spring unstretched. Find the deflection under the load *P* in terms of *P*, *E*, *I*, *L*, and the spring constant *k*.

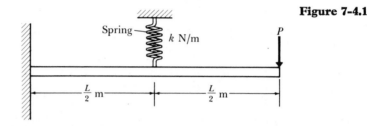

Figure 7-4.1

7-4.2 Determine the reactions of the following *L*-shaped bar in terms of *P*, *E*, and *I*. Assume that the cross section of the bar is circular and that the bar is not rigidly attached but simply rests on the block at *B*.

Figure 7-4.2

$$EI = 1.5GJ = \text{const.}$$

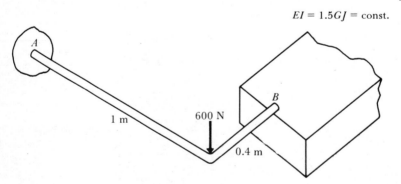

7-4.3 The beam *AC* is built into a rigid body at end *A*, and shaft *ED* is fixed against rotation at *E*. The relatively rigid metal arm is shrunk on the shaft at *D*. The end of the arm and the end of the beam are just in contact at *C* when there is no load on the beam. A load *P* of 120 lb is applied to the beam as shown.

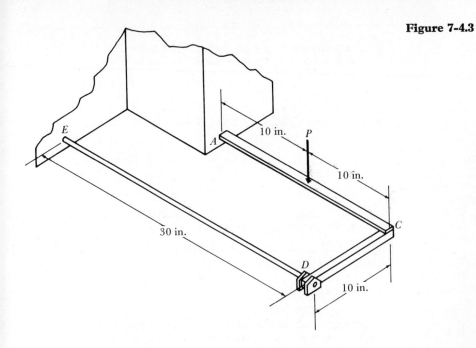

Figure 7-4.3

Beam AC is of aluminum alloy, 2 in. wide and $\frac{1}{2}$ in. thick ($E = 11 \times 10^6$ psi). The shaft ED is made of steel ($G = 11.6 \times 10^6$), 0.6 in. in diameter. Calculate the maximum normal stress in the beam. Also calculate the maximum shearing stress in the shaft.

7-4.4 The W 8 × 31 beam is rigidly welded to a hollow pipe at each end. Both the pipe and the beam are of structural steel. The pipe has a circular cross section with a 7-in. outer diameter and a $\frac{1}{4}$-in. wall thickness. Determine the reactions at A and B. The pipe ends are restrained from rotation.

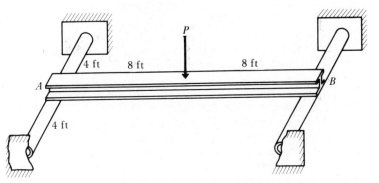

Figure 7-4.4

Chapter eight

Inelastic action

R8-1 Plastic deformation of metals, strain-hardening, temperature effects, Bauschinger's effect, fatigue, creep

When we speak of elastic action of a material, we are referring to a distortion on the atomic and/or molecular level produced by the action of external forces. During elastic action, the atoms and molecules maintain their relative positions while the forces are applied, and will return to their original, undistorted configuration after the forces are removed.

A material that undergoes stress beyond a certain limiting stress, designated as its elastic limit σ_e, will break (fracture) if it is an ideally elastic (or brittle) material, or yield if it is a ductile material. Figure 8-1 shows an idealized σ–ϵ diagram for both a brittle material and a ductile material. Note that, in this particular example, both materials can theoretically sustain the same service load, but the ductile specimen has much greater toughness (area under σ–ϵ curve), which means that it is able to absorb much more energy before fracture. We shall have more to say about this later.

There are some materials, such as glass, which have a σ–ϵ diagram similar to that shown for the brittle specimen in Figure 8-1, but even glass yields a small amount before fracture. Since most engineering materials will yield an appreciable amount before fracturing, let us examine the nature of yielding more closely.

When a material is subjected to stresses that are larger than its elastic limit σ_e, the distances between atoms increase to such an extent that atomic bonds are broken, and place changes occur in the atomic lattice of individual crystals. Since these place changes are changes of positions of atoms relative to one another, the result is a much larger overall deformation of the material, and the action is not reversible when the external force is relaxed or removed. This nonrecoverable deformation is called *plastic deformation*.

Remember, then, that plastic deformation or yielding of a material is primarily a rearranging of atoms along certain slip planes. Yielding in a metal does not result in any loss of strength, because the new atomic bonds are just as strong as the old ones.

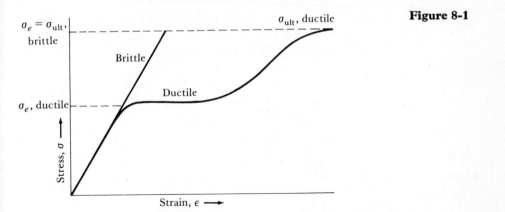

Figure 8-1

Slip is the most common mechanism by which plastic deformation occurs in a single crystal. It refers to a large displacement of one part of the crystal with respect to another along certain well-defined planes within the crystal. The crystallographic planes along which slip occurs are called the *slip planes*. Slip takes place in a crystal when the shearing stress on the slip plane reaches a critical value associated with that crystal. This critical value is a characteristic quantity of each material, and varies from material to material. If the critical shearing stress that initiated the yielding process is held constant, yielding will soon stop, and will resume only when the shearing stress is increased to a higher value. This increase in the value of critical shearing stress after initial yielding is called *strain-hardening*.

It is possible to obtain a theoretical estimate of the critical shearing stress in a crystal by calculating the cohesion of a solid in terms of the energy required to decompose it into isolated atoms or molecules. Experimental evidence shows, however, that the observed critical shearing stress is much lower (hundreds of times lower) than the calculated one. The reason for the discrepancy is that the calculations are based on the assumption of "perfect" crystals, while actual crystals contain many imperfections that allow slips to start at relatively low stresses.

Practically all engineering materials are polycrystalline in nature, i.e., they are composed of many crystals, usually randomly oriented. In such a material, the mechanism that produces plastic deformation is more complicated, because of the effect of neighboring crystals and the interference of the crystalline or grain boundaries. It is sufficient for our purposes to mention simply that the effect of the grain boundaries is to increase the elastic strength and, to a lesser degree, the plastic strength of a material, because the boundaries interfere with the slip across the individual crystals. It follows, then, that a polycrystalline metal has higher strength than a single crystal, and that fine-grained metals have higher strength than coarse-grained ones, because of the larger number of grain boundaries in the fine-grained metals.

Figure 8-2 illustrates the strain-hardening effect in a polycrystalline metal as observed during a typical cyclic loading–unloading tensile test on a specimen of mild (low-carbon) steel.

Figure 8-2

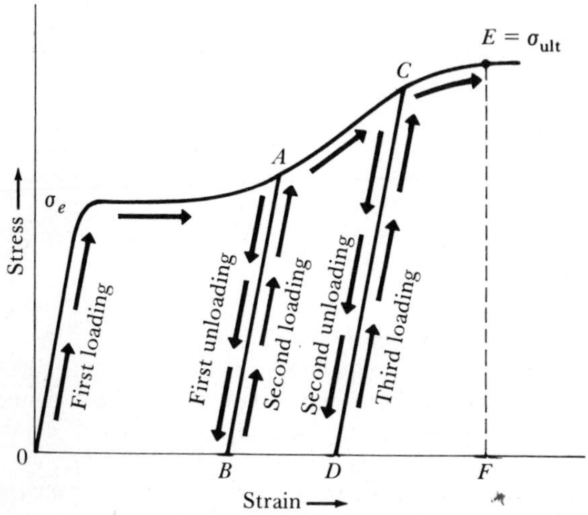

First, consider how yielding starts and continues in such a specimen. It is reasonable to assume that yielding will start in a region of high stress, because a truly uniformly distributed stress is extremely difficult to achieve in a real member of an engineering material. Thus a concentration of stress may be caused by the method of loading, the geometry of the member, material imperfections, such as inclusions of foreign matter or voids, etc. In a tension test such as that illustrated in Figure 8-2, the highest stress concentration will probably occur near the grips of the testing machine. Slip will first occur in those crystals whose slip planes are most favorably oriented in this area of high stress. When slip occurs, a rotation of the crystal with respect to the axis of loading will also take place. The rotation will bring other slip planes into a more favorable position, and since the length of a slip line may be a thousand times the distance between atoms, other crystals in the neighborhood are now subjected to more stress and begin yielding. This process repeats itself until all fibers are in the plastic range. What we actually observe while the specimen yields is the cumulative effect of a great many macroscopic, microscopic, and submicroscopic (atomic) actions.

Now observe what happens to our tension specimen if we first load it beyond the yield point, but not to fracture (point A in Figure 8-2), then unload and reload it again, as indicated by the arrows in Figure 8-2. If the loading and unloading process is carried out at a slow rate and at room temperature, the unloading curve *AB* will be approximately parallel to the straight-line portion of the loading curve *OA*.

Reloading the specimen now again in tension, we find that the new diagram retraces the curve from *B* back to *A* and then continues toward *C*. An inspection of the diagram will indicate that the elastic limit of our specimen has been raised for the second loading as a result of the first loading into the plastic range. The raising of the elastic limit by this action is also called *strain-hardening*, or sometimes *work-hardening*. Note that the second loading produces essentially the same curve *AC* beyond the new elastic limit that would be followed by an uninterrupted first loading.

This process may be repeated to raise the elastic limit even higher, as shown by the curve from *C* to *D* to *C* and beyond. This method of strain- or work-hardening may thus be used successfully to increase the elastic range of a material, but note that the ultimate strength *E* remains essentially unchanged.

It is also important to note here that, unfortunately, everything is not as favorable as it appears to be. You may recall that toughness, i.e., the ability of a material to absorb energy, is measured by the total area under the stress–strain diagram. An examination of Figure 8-2 shows that the area under the stress–strain diagram of the strain-hardened specimen (area *BACEF*) is much smaller than that of the non-strained-hardened one (area *OACEF*). Thus we see that strain-hardening leaves the material in a more brittle state (less ductility and less toughness). It is also true that the elastic strength will not increase if the direction of loading is reversed. With many metals, exceeding the elastic limit in tension lowers the elastic limit in compression. This phenomenon, called the *Bauschinger effect*, is illustrated in Figure 8-3.

Figure 8-3

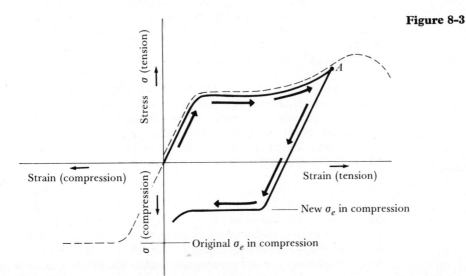

The dashed line in Figure 8-3 indicates the idealized behavior of a metal subjected to tension and compression tests. The solid lines show what happens when a specimen of this metal is first strained to point *A* in tension and then unloaded and subjected to a compression force. Note that the straight-line portion of the compression stress–strain diagram is shorter for the specimen that was first strain–hardened in tension than for the original non-strain-hardened metal.

We already mentioned earlier that another name given to the phenomenon of strain-hardening is work hardening. A third name which refers to the same phenomenon is *cold working*. This expression is more descriptive, because it emphasizes that the plastic deformations in the material are produced at temperatures below the recrystallization temperature of the material. That is, they are produced at a temperature which is below that at which the formation of new grains and the dissolution of old grains can readily take place. Any kind of mechanical action, such as bending, twisting, shot peening, pulling, etc., resulting in a permanent change of shape of a metal at room temperature can be classified as cold working. The result of cold working of metals is usually an increase in hardness and elastic strength of the material, but the effect is accompanied by a decrease in ductility and toughness.

After cold working, the metal is in a high-energy state, because part of the work done which produced the plastic deformation is stored as internal energy. The natural tendency of the material is to return to a lower-energy state. At room temperature this is an extremely slow process for most engineering materials. Notable exceptions are lead and tin, whose recrystallization temperatures are below room temperature. Consequently, these materials cannot be work-hardened at room temperature. *Annealing* is the name given to the process of reversing the effects of cold working on metals by heating and subsequent cooling under carefully controlled conditions. In general, fine-grained metals have higher strength and less ductility than coarse-grained ones. A slow rate of cooling from above the recrystallization temperature of a metal will produce larger grains, while a fast rate of cooling will produce finer grains.

Hot working is the term applied to the process of shaping metals at elevated temperatures (usually just above the recrystallization temperature). This process is more expensive than cold working, but it ensures that the mechanical properties of the material are not affected by the plastic deformation. Hot working actually represents a combination of cold work and simultaneous annealing.

When a material is subjected to many cycles of loading and unloading, the maximum stress required for fracture is less than is required under a steady application of load. The engineering term applied to this type of effect or failure is *fatigue*. Fatigue is by far the most common reason for fracture of moving parts, such as can be found, for example, in motors, turbines, pumps, and similar machinery. You have probably made use of the tendency of materials to break more readily under repeated loads when you wanted to break a wire. You probably broke it by bending it back and forth several times. If you never had occasion to do this before, take a paper clip, open it up, and bend it back and forth several times. You will observe that, even though the material seems to be in no way damaged after a few bending cycles, it will suddenly break. As a matter of fact, the material behaves as though it finally "got tired" or "fatigued" under the repeated stressing and unstressing.

The exact reason for fatigue failure is quite complex and not completely understood. A simple description of the mechanism of fatigue is that a very small crack originates at a point of high stress. This point may be at some imperfection inside the material or at a surface scratch or at some abrupt change of geometry, such as a sharp corner, all of which may lead to stress concentrations (see Section R8-2). The crack enlarges during the continued repetitive loadings until there is not enough undamaged material left to support the load. The resulting failure is often sudden and catastrophic because even so-called ductile materials fail in an essentially brittle fashion, i.e., with no noticeable yielding to warn of impending failure.

A simple uniaxial tension or torsion test does not provide us with sufficient information for designing a member that may be subjected to repeated loadings which may result in fatigue failure. Special tests must be conducted to determine a material's susceptibility to fatigue failure.

One of the most commonly used tests for this purpose is the rotating-beam test. In this test, the number of completely reversed cycles of bending stress required to cause failure is determined for various stress levels. The test data are usually plotted on semilog paper, and result in graphs similar to those shown in Figure 8-4. These graphs are usually called *S–N curves*.

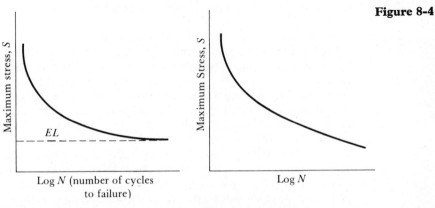

Figure 8-4

(a) Ferrous metal   (b) Nonferrous metal

Figure 8-4(a) shows a typical *S–N* curve for a ferrous metal, such as steel. Note that for this type of metal there exists a stress level (marked *EL*) below which failure will not occur regardless of the number of stress cycles. This stress is called the *endurance limit* or *fatigue limit* of the material, and is generally about one-half the ultimate strength.

Figure 8-4(b) shows a typical *S–N* curve for a nonferrous metal, such as aluminum. This curve does not show an endurance or fatigue limit. Since this limit does not always exist, the terms *endurance strength* or *fatigue strength* are more commonly used to define the strength of materials under repeated stresses. The endurance or fatigue strength of a material is defined as the maximum stress that can be applied repeatedly over a specified number of stress cycles without producing fracture of the material. The fatigue strength of a material diminishes at elevated temperatures.

Creep is yet another phenomenon that can result in failure of a material at stress levels insufficient to cause fracture under conditions of a uniaxial tension test. Creep is defined as the slow and progressive deformation, over time, of a material under constant load. If creep is allowed to take place over a sufficient length of time, it may result in fracture of the material. A few metals, lead for example, will exhibit creep at room temperature, but most metals will creep only at higher temperatures. Creep tests are difficult to perform, because the creep phenomenon is highly dependent on temperature and time.

Turn to the Study Guide. ■ STOP

R8-2 Stress concentrations

The stress formulas as derived in previous chapters will give stress values that are radically different from actual values at points in a member where abrupt changes in section occur, such as around a rivet hole in an axially loaded plate (Figure 8-5).

Figure 8-5(a) is a free-body diagram of an axially loaded plate of width b and thickness t. If we investigate the stresses at section *aa*, we will find them to be very close to the values predicted by the formula $\sigma = P/A$. Figure 8-5(b) is another free-body diagram of the same plate, but here the stresses in question exist at a section cut through the rivet hole. The dashed lines indicate the magnitude and distribution of the stresses as predicted by the formula $\sigma_n = P/A_n$, with A_n representing the net cross-sectional area $[A_n = (b - 2\rho) \cdot (t)]$. The solid lines show the actual magnitude and distribution of the stresses on this section. Note that the stress at the edge of the hole is significantly larger than P/A_n. The factor k by which we must multiply P/A_n to get the actual stress is called the *stress-concentration factor*. The stress-concentration factor for this plate will approach 3 if the width b is very large and the radius of the hole ρ is very small. (If the diameter of the hole is increased to about one-half the value of the plate width b, the stress-concentration factor will decrease to about 2.1.)

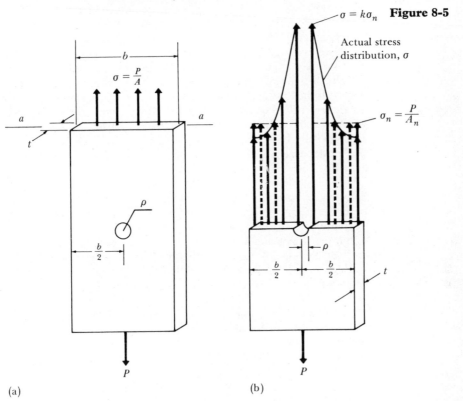

(a)

(b)

The actual stress distribution near a small hole in a very wide, axially loaded plate can be determined mathematically by the theory of elasticity. In many members, however, a mathematical analysis of the actual magnitude and distribution of stress across sections containing discontinuities is very difficult, and in some cases impossible. This is why experimental methods of stress analysis are often used to determine stress-concentration factors. A complete treatment of the theory of elasticity and experimental stress analysis is beyond the scope of this text. However, the following brief descriptions will help you understand what the subject is all about.

The theory of elasticity differs from the ordinary mechanics of materials in the assumptions in the respective methods concerning the distribution of strains in a body. Both methods assume that the material of which the member is made is homogeneous and isotropic, and that there is a definite relationship between stress and strain as expressed by Hooke's law, and both methods satisfy the conditions of equilibrium. However, in ordinary mechanics, we make assumptions concerning the overall distribution of strains in the body as a whole, while in the theory of elasticity, we require the deformations of each element of the body to obey the mathematically formulated law of elastic continuity of the material. This means that, in the theory of elasticity, we make no simplifying assumption concerning the strain, so that it is necessary to use the general statement of Hooke's law to express the relationship between stresses and strains. Because of the more exact mathematical formulation of problems in the theory of elasticity, its method is sometimes referred to as the *exact* method. It must be kept in mind, however, that either method is only as exact as the original assumption of an ideal homogeneous, isotropic, linearly elastic material. Since no actual material can completely fulfill these requirements, both methods will yield only approximate results. However, they are both of great value and supplement each other as methods for determining the magnitude and distribution of stresses in actual bodies.

The assumptions made in ordinary mechanics are usually based on observation and measured strains in bodies subjected to loads, or in models of these bodies. Some of the more common methods of determining strains are (1) elastic-strain or strain-gage method, (2) photoelastic analysis, (3) brittle-coating method, (4) rubber models, (5) membrane or soap-film analogy. All of these and other methods are discussed in detail in any text on experimental stress analysis.

Now let us return to our discussion of the axially loaded plate of Figure 8-5. We shall assume that the plate is made of a ductile material whose stress–strain diagram can be approximated by that shown in Figure 8-6(a). (This assumption disregards the effects of strain-hardening that take place beyond the elastic limit of the material. Stresses that are larger than the elastic limit σ_e are considered "emergency reserve" strength of the material, not "usable-design" strength.) Figure 8-6(b) shows the same section as Figure 8-5(b), but the external load P has now been increased to a value that will produce nominal stresses $\sigma_n = P/A_n$ which are close to the elastic-limit stress σ_e of the material. Multiplying this stress by the stress-concentration factor k yields stresses near the hole that are above the elastic limit of the material (see the dashed lines in Figure 8-6(b). Since it is impossible for this material to develop stresses above the elastic limit, the actual stress distribution across this section is similar to that shown by the solid lines in Figure 8-6(b).

Only a so-called ductile material, such as low-carbon steel, which has a stress–strain curve like that shown in Figure 8-6(a), is able to develop the stress distribution shown by the solid lines in Figure 8-6(b). A brittle material, such as glass, cast iron, or brick, cannot redistribute stresses in this fashion; instead, a small crack would appear at the most highly stressed point as soon as the elastic limit of the material was reached (recall that the elastic limit and the ultimate strength of very brittle materials are the same). A more detailed discussion of stress redistribution and residual stresses will be given in the Study Guide.

Small holes in an axially loaded plate are not the only conditions that will produce stress-raising effects. Any place in a structure at

which an abrupt change in geometry occurs is a potential danger
point, and stresses in the vicinity of such discontinuities may be many
times larger than those predicted by our basic stress equations.

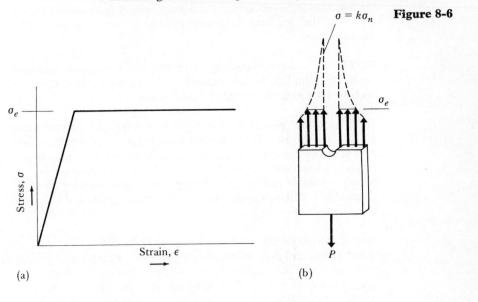

Figure 8-6

(a)

(b)

Turn to the Study Guide. ■ **STOP**

R8-3 Inelastic action: torsion

In Chapter 3, we derived the important torsional-shear-stress equation

$$\tau = \frac{T\rho}{J} \tag{1}$$

This equation is valid only for linear elastic action. If we want to
know the relationship between torque and shearing stress beyond the
linear elastic range, we must go back to the integral relationship (see
Chapter 3):

$$T = \iint\limits_{\text{Area}} \tau_\rho \rho \; dA \tag{2}$$

which depends only on statics for its validity, and is therefore valid
for states of elastic as well as inelastic stress.

In order to evaluate Equation (2) for torque on a given cross section,
we must know how the shearing stress τ_ρ varies as a function of the
radial distance ρ. This information is usually obtained from (a) the
shearing-strain distribution in the torsion member and (b) the relationship between shearing stress and shearing strain for the material.

In the analysis of plastic torsion members, we make the same assumptions that were made for elastic circular torsion members (except for the assumption of linear elastic action). These assumptions lead us to the conclusion that shearing strains vary linearly from zero at the center of the specimen to a maximum value at the outside of the specimen.

▶ **NOTE** We can experimentally verify this conclusion that shearing strain varies linearly from the center to the outside surface of the specimen for stresses beyond the elastic range.

If we know the value of the shearing strain at one point on the cross section, and we assume that the shearing strain varies linearly from zero at the center of the specimen outward, we can determine the strain at all radial distances. The only additional information we need to evaluate the integral relating torque and shearing stress (Equation 2) is the shearing-stress-versus-shearing-strain relationship for the material.

Most materials exhibit work-hardening when loaded beyond the linear elastic range, and their shearing-stress-versus-shearing-strain relationship may be quite complex. To simplify calculations, stress–strain diagrams are often approximated by straight-line segments. For example, a material with a stress–strain diagram as shown in Figure 8-7 could be approximated by two straight-line segments. The first straight-line segment represents the linear elastic action, and the second segment the work-hardening phase.

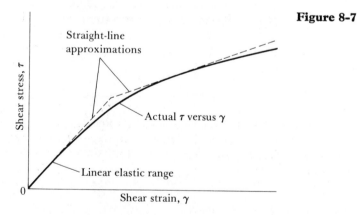

Figure 8-7

Some materials exhibit little or no work-hardening effects beyond the linear elastic range. Mild steel, for example, shows practically no work-hardening effects until the strains become ten to twenty times the strain at the proportional limit. Materials which exhibit little or no work-hardening can be approximated by a linear elastic stress–strain curve followed by an ideally (fully) plastic stress–strain curve, as shown in Figure 8-8. The values τ_e and γ_e represent respectively the shearing-stress and shearing-strain limits of linear elastic action.

Figure 8-8

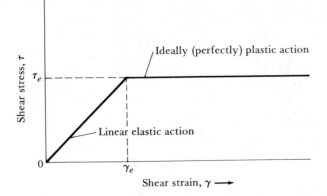

NOTE The terms yield point, elastic limit, and proportional limit all denote the same point on the idealized stress–strain diagram. The stresses and strains at this point are identified by the subscript e (for example, τ_e, σ_e, γ_e, ϵ_e).

A material whose stress–strain curve is approximated by that shown in Figure 8-8 is sometimes called an *elastoplastic* material (or elastic and perfectly plastic material).

An elastoplastic circular torsion member will have one of the three stress states shown in Figure 8-9. If the strain does not exceed the yield point, the action is linearly elastic, as shown in Figure 8-9(a). After the shearing strain exceeds the yield point γ_e, the member becomes elastic over a portion of the cross section to some radial distance a, and the stress beyond the elastic range is the yield-point stress τ_e, as shown in Figure 8-9(b). In the limit (as the distance a becomes very small), we can approach a condition known as *fully plastic torsion*, as illustrated in Figure 8-9(c). The fully plastic torque is the largest torque that the specimen can carry, unless the strains become tremendously large and work-hardening occurs.

In general, to evaluate the integral for the torque (Equation 2) of elastoplastic torsion members, we have to evaluate two integrals, since τ_ρ is sectionally continuous. One integral will cover the elastic core, and the second will cover the plastic region beyond the elastic core. The integrals for a solid shaft that is partially elastic and partially plastic (Figure 8-9b) are as follows.

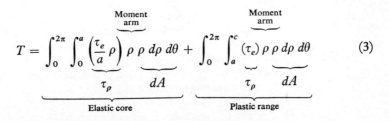

$$T = \int_0^{2\pi} \int_0^a \left(\frac{\tau_e}{a}\rho\right) \rho\, \rho\, d\rho\, d\theta + \int_0^{2\pi} \int_a^c (\tau_e)\, \rho\, \rho\, d\rho\, d\theta \qquad (3)$$

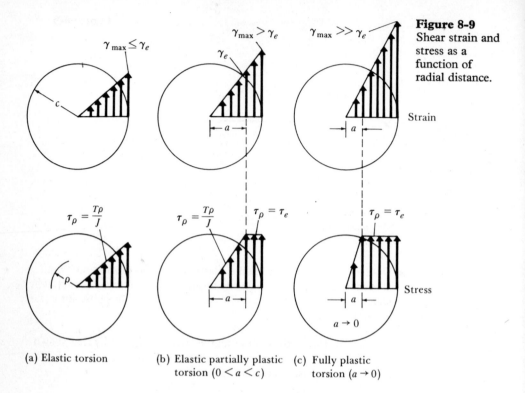

Figure 8-9
Shear strain and stress as a function of radial distance.

(a) Elastic torsion

(b) Elastic partially plastic torsion $(0 < a < c)$

(c) Fully plastic torsion $(a \to 0)$

The limiting elastic case (maximum elastic torque) occurs when $a \to c$ and the second integral drops out of Equation (3). When $a \to 0$, thus eliminating the first integral in Equation (3) and leaving us with the second integral, we have an approximation of the fully plastic torque.

Ordinary carbon steels are frequently approximated by linear elastic and ideally plastic stress–strain curves. The ratio of the fully plastic torque to the maximum elastic torque is used as a measure of the increased torque-carrying capacity that can be obtained by designing in the plastic range.

The assumption that the strain varies linearly from zero at the center of the specimen outward permits us to determine the angle of twist (see Chapter 3) from the relationship

$$\gamma = \frac{\rho\theta}{L} \quad \text{or} \quad \gamma_c = \frac{c\theta}{L} \tag{4}$$

In summary: If we know the shearing-stress-versus-shearing-strain relationship for a material, as well as the strain at any radial distance, we can assume a linear strain distribution and determine the torque and angle of twist by means of Equations (2) and (4), respectively. Conversely, we can determine the strain and stress distribution from known values of torque or angle of twist.

Turn to the study program in Section SG8-3. ■ STOP

R8-4   Inelastic action: bending

In Chapter 4, we derived the flexure-stress formula

$$\sigma = \frac{My}{I} \tag{5}$$

The validity of the formula depends on several assumptions, including the assumption of linear elastic action.

The normal-stress-versus-normal-strain relationship for most materials is quite similar to the shearing-stress-versus-shearing-strain relationship. There is generally a linear elastic range followed by a strain-hardening range.

Some materials, notably ordinary carbon steels, exhibit little or no strain-hardening until the strains become ten to twenty times the strain at the proportional limit (or yield point). These materials have a stress–strain curve that can be approximated by an elastoplastic curve as defined in Section R8-3.

In the analysis of bending beyond the linear elastic range, we shall make the following assumptions.

1 The beam cross section must have an axis of symmetry parallel to the loading direction.
2 Cross sections remain plane after bending.
3 The beam must be straight, with a constant cross section.
4 The material is homogeneous and isotropic.

Similar assumptions were made in the analysis of elastic flexural members, and led to the same conclusion: The normal strains are zero at the neutral axis of the cross section, and they vary linearly with the perpendicular distance from the neutral axis. Therefore

$$\epsilon = ky, \quad \text{where } k = \epsilon_c/c \tag{6}$$
$$\text{and } y = \text{distance from the neutral axis}$$

$c = $ maximum distance from neutral axis

$\epsilon_c = $ strain at distance c

▶ **NOTE** We can experimentally verify that the strain with values considerably beyond the range of linear elastic action varies linearly from the neutral axis.

The bending moment in a flexurally loaded member is given by the following integral (Chapter 4):

$$M = \iint_{\text{Area}} \sigma y \, dA \tag{7}$$

The integral for bending moment depends only on statics for its validity. Thus it is valid for inelastic as well as elastic stress states.

In order to evaluate the integral for the bending moment on a given cross section, we need to know how the normal stress σ varies as a function of the distance y from the neutral axis. The stress distribution can be obtained from the normal-strain distribution in the member, and the normal-stress-versus-normal-strain relationship for the material. In addition, we also need to know the location of the neutral axis of bending.

▶ **NOTE**   Compare the bending-moment integral (Equation 7) and the integral for torque [Equation (2) of Section R8-3]. Both integrals involve stress, moment arm, and area. Both are integrals for (couples) moments. The terms "torque" and "bending moment" simply refer to how the moment is applied relative to the geometry of the member.

If the member is subjected to bending only (no axial loads), then the summation of forces on the cross section must equal zero (i.e., the stress distribution on the cross section must give a couple). This fact allows us to locate the neutral axis on the basis of the following equation:

$$\sum F_{\text{normal}} = \iint_{\text{Area}} \sigma \, dA = 0 \tag{8}$$

● **CAUTION**   The neutral axis of bending is not necessarily at the centroid of the cross section. Its location depends on the stress distribution as a function of y, and the shape of the cross section.

In this section, we shall consider only ideally elastoplastic materials. We shall further assume that the materials have the same stress-strain relationships and yield points in tension and compression.

In general, if the cross-sectional area is not symmetrical with respect to the centroidal axis, the neutral axis will move away from the centroid as the yield zone progresses. For a fully plastic bending moment (plastic hinge), the normal tensile and compressive stresses on the cross section will be of equal magnitude, and the neutral axis will be the axis that divides the cross-sectional area equally. In the case of symmetry about the centroidal axis, the neutral axis will always be the centroidal axis.

Consider the case of a rectangular member subjected to bending loads. One of the three stress distributions in Figure 8-10 will be applicable if the material is elastoplastic. Since the cross section and the stress distribution are symmetric with respect to the centroidal axis for the rectangular section of Figure 8-10, the neutral axis for plastic action will remain at the centroid.

To evaluate Equation (7) for partially plastic action, we need at least two integrals, since the stress is sectionally continuous.

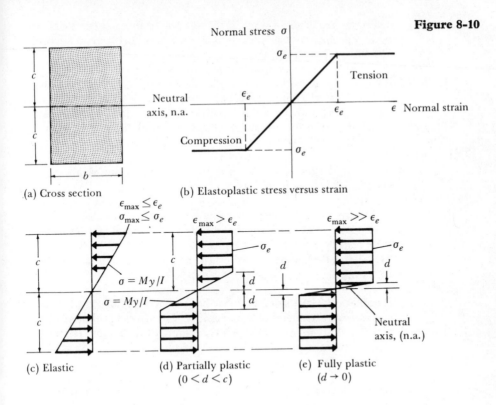

(a) Cross section

(b) Elastoplastic stress versus strain

Figure 8-10

(c) Elastic

(d) Partially plastic
$(0 < d < c)$

(e) Fully plastic
$(d \to 0)$

Turn to the study program in Section SG8-4. ∎ **STOP**

Problems

8-3.1 Determine the ratio of the fully plastic torque to the maximum elastic torque for a hollow circular shaft whose inside diameter is one-half the outside diameter. Assume that the material is elastoplastic.

8-3.2 Determine the percentage reduction in the capacity of a solid circular shaft to carry fully plastic torque, given that one-half of the cross section is removed by drilling an axial hole through the length of the specimen. Note that the drilling corresponds to a 50% reduction in weight.

8-3.3 A solid circular shaft is made of a material that can be idealized as an elastoplastic material with a shearing yield point of 40,000 psi. The shaft is 8 ft long and has a radius of 2.5 in. Determine the shearing-stress distribution on the cross section when the shaft carries a torque of 92,000 lb-ft. Also determine the angle of twist of the shaft. ($G = 10 \times 10^6$ psi)

8-3.4 A solid steel shaft 100 mm in diameter carries a torque of 31 kN·m. Determine the radial distance at which plastic action starts. Also determine the angle of twist, given that the member is 3 m long. Assume that the material is elastoplastic, with a shearing yield point of 126 MPa. ($G = 80$ GPa)

8-3.5 A solid circular shaft of 2-in. radius can be approximated as an elastoplastic material with a shearing yield point of 20,000 psi. What torque can this shaft carry if the yield zone penetrates 1.0 in.; that is, the elastic zone goes from the center to 1.0 in. from the center. ($G = 12 \times 10^6$ psi)

8-3.6 An elastoplastic solid circular steel shaft has a shearing yield point of 154 MPa. ($G = 80$ GPa)

a) Determine the shearing strain at the surface of the shaft when the shearing stress at three-eighths of the radius from the center of the shaft reaches the yield point.
b) What is the torque carried by the shaft of part (a) if the radius is 25 mm?

8-3.7 A hollow circular torsion member has an outside diameter of 4 in. and an inside diameter of 3 in. Assume the material is elastoplastic, with a yield point of 12,000 psi. Determine the maximum elastic torque, the fully plastic torque, and the ratio of T_{fp}/T_e. ($G = 12 \times 10^6$ psi)

8-3.8 An elastoplastic hollow steel shaft has an outside diameter of 100 mm and an inside diameter of 50 mm. The shearing yield point is 140 MPa. Determine the fully plastic torque, the maximum elastic torque, and the ratio T_{fp}/T_e. ($G = 80$ GPa)

8-4.1 Determine the ratio of the fully plastic bending moment to the maximum elastic bending moment for a member with a solid circular cross section.

8-4.2 Determine the ratio of the fully plastic bending moment to the maximum elastic bending moment for a member with a rectangular cross section of base b and height h.

8-4.3 Determine the fully plastic bending moment, the maximum elastic bending moment, and the ratio M_{fp}/M_e for the cross section shown. The material is elastoplastic, with a yield point of 30 ksi. ($E = 30 \times 10^6$ psi)

8-4.4 A beam is fabricated from three 20 mm × 120 mm steel plates to form an I section, as shown. Given that the material is elastoplastic, with a yield point of 280 MPa, what are the fully plastic bending moment and the maximum elastic bending moment for the section? Also determine the ratio M_{fp}/M_e.

8-4.5 Determine the ratio of the fully plastic bending moment to the maximum elastic bending moment for a W 14 × 87 standard structural-steel member. Assume that the material is elastoplastic. Neglect the effects of fillets and rounds in your calculations.

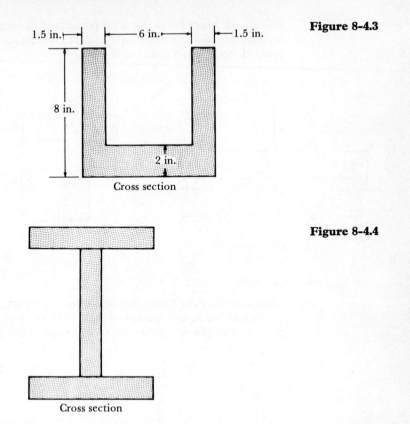

Figure 8-4.3

1.5 in.

6 in.

1.5 in.

8 in.

2 in.

Cross section

Figure 8-4.4

Cross section

8-4.6 Determine the ratio of the fully plastic bending moment to the maximum elastic bending moment (M_{fp}/M_e) for the cross section shown. Assume that the material is elastoplastic.

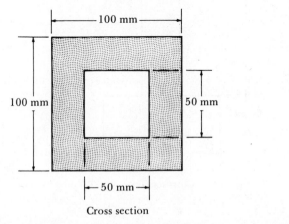

Figure 8-4.6

100 mm

100 mm

50 mm

50 mm

Cross section

8-4.7 A simply supported beam 6 ft long is loaded at its center with a concentrated load P. Determine the load P that will cause the beam to develop a plastic hinge (fully plastic bending moment). When the fully plastic hinge forms, for what portions of the beam will the flexural stresses still be entirely in the linear elastic range? The beam is made of steel, with a yield point of 30,000 psi.

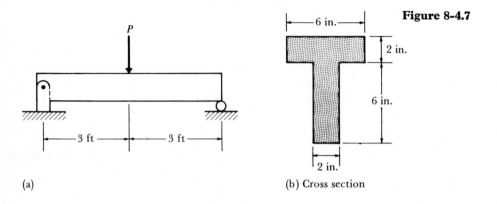

Figure 8-4.7

(a)

(b) Cross section

8-4.8 Determine the fully plastic bending moment, the maximum elastic bending moment, and the ratio M_{fp}/M_e for the cross section shown. Assume elastoplastic action with a yield point of 210 MPa.

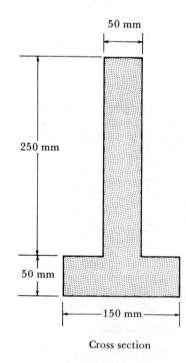

Figure 8-4.8

Cross section

Chapter nine

Energy method

R9-1 Deflections of statically determinate elementary structures by Castigliano's method

The method that we are going to discuss in this chapter involves a quantity called strain energy. First we shall define strain energy and learn how to calculate it for a given structure. Then we shall use a method called Castigliano's method to determine deflections of chosen points on a given structure.

Strain energy   If we apply external forces to a body, the body deforms and work is done by the forces. Do you remember the first law of thermodynamics? It is

$$W_e + Q = \Delta T + \Delta U \qquad (1)$$

where

W_e = the work done on the body by external forces
Q = the heat that flows into the body
ΔT = the increase in kinetic energy
ΔU = the increase in internal energy

If we assume that no heat exchange takes place (adiabatic process) and the external loads are applied slowly, then all the work of the external forces is stored within the body in the form of internal energy, and Equation (1) becomes

$$W_e = U \qquad (2)$$

▶ **NOTE** Equation (2) implies that no energy is dissipated and the work done by the external forces is recoverable internal energy. Thus the material behaves in an ideally elastic manner.

The quantity ΔU, which is the change in internal energy of the material, exists because the material has been deformed or strained under the action of the external forces. Thus U is usually called *strain energy*.

Let's determine the strain energy of a rectangular element of a material that is subjected to a normal stress σ_x, as shown in Figure 9-1(a). Figure 9-1(b) shows the strained configuration of the element.

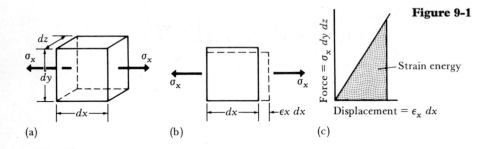

Figure 9-1

(a)   (b)   (c)

▶ **NOTE**   The relationship between force $(\sigma_x\,dy\,dz)$ and displacement $(\epsilon_x\,dx)$ is shown in Figure 9-1(c) as a straight line, where we assume that the material behaves linearly according to Hooke's law.

The work done on the element, or the strain energy absorbed by the element, is equal to the area under the force–displacement curve of Figure 9-1(c). Thus

$$dU = \underbrace{\tfrac{1}{2}\sigma_x\,dy\,dz}_{\text{Force}}\ \underbrace{\epsilon_x\,dx}_{\text{Displacement}} \tag{3}$$

or

$$dU = \tfrac{1}{2}\sigma_x\epsilon_x\,dx\,dy\,dz \tag{4}$$

Since the volume of the element (dv) is equal to $dx\,dy\,dz$, Equation (4) becomes

$$dU = \tfrac{1}{2}\sigma_x\epsilon_x\,dv \tag{5}$$

The strain energy of the entire body arising from the normal stress σ_x is obtained by integrating Equation (5):

$$U = \tfrac{1}{2}\int_v \sigma_x\epsilon_x\,dv \tag{6}$$

where v is the volume of the entire body.

Now let's determine the strain energy of the same rectangular element of material shown in Figure 9-1 when it is subjected to the shear stresses shown in Figure 9-2(a).

Figure 9-2(b) shows the deformed configuration of the element as a result of the shear stresses. In Figure 9-2(b), the bottom edge is fixed while the other three edges are allowed to move; however, this restriction causes no loss in generality. The forces $\tau_{xy}\,dy\,dz$ on the vertical faces of the element do no work, since they are perpendicular to the displacement $\gamma_{xy}\, dy$. The work done by the force $\tau_{xy}\, dx\, dz$ on the top face is

$$\underbrace{\tfrac{1}{2}(\tau_{xy}\, dx\, dz)}_{\text{Force}}\ \underbrace{(\gamma_{xy}\, dy)}_{\text{Displacement}}$$

if the material behaves linearly. The work done by the force on the bottom face of the element is also zero, since it does not move through a displacement. The total work on the element (or the strain energy absorbed by the element) is therefore

$$dU = \tfrac{1}{2}\tau_{xy}\gamma_{xy}\, dx\, dy\, dz$$

or

$$dU = \tfrac{1}{2}\tau_{xy}\gamma_{xy}\, dv \tag{7}$$

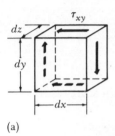

(a)

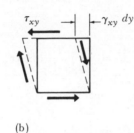

(b)

Figure 9-2

The strain energy absorbed by the entire body from the shear stress τ_{xy} is obtained by integrating Equation (7) over the entire volume:

$$U = \tfrac{1}{2}\int_v \tau_{xy}\gamma_{xy}\, dv \tag{8}$$

If we add up the strain energy absorbed by the element from the six stress components of a three-dimensional state of stress, we obtain

$$dU = \tfrac{1}{2}(\sigma_x\epsilon_x + \sigma_y\epsilon_y + \sigma_z\epsilon_z + \tau_{xy}\gamma_{xy} + \tau_{xz}\gamma_{xz} + \tau_{yz}\gamma_{yz})\, dv \tag{9}$$

The general expression for the strain energy absorbed by the entire body, which is obtained by integration of Equation (9) over the entire volume, is

$$U = \tfrac{1}{2}\int_v (\sigma_x\epsilon_x + \sigma_y\epsilon_y + \sigma_z\epsilon_z + \tau_{xy}\gamma_{xy} + \tau_{xz}\gamma_{xz} + \tau_{yz}\gamma_{yz})\, dv \tag{10}$$

Strain energy: centric loading For the centrically loaded member shown in Figure 9-3, the six stress components at any point in the member are

$$\sigma_x = \frac{P}{A}, \qquad \sigma_y = \sigma_z = \tau_{xy} = \tau_{xz} = \tau_{yz} = 0$$

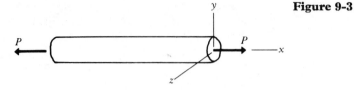

Figure 9-3

The strain energy, according to Equation (10), is

$$U = \tfrac{1}{2} \int_v \sigma_x \epsilon_x \, dv$$

Since the material must be linearly elastic, Hooke's law applies ($\epsilon_x = \sigma_x/E$), so that the strain energy becomes

$$U = \frac{1}{2} \int_v \sigma_x \left(\frac{\sigma_x}{E}\right) dv = \frac{1}{2E} \int_v \sigma_x^2 \, dv$$

or

$$U = \frac{1}{2E} \int_v \frac{P^2}{A^2} \, dv = \frac{1}{2E} \int_v \frac{P^2}{A^2} \, dA \, dx$$

Since the cross-sectional area is constant, the strain energy is

$$U = \frac{1}{2EA^2} \int_L P^2 \, dx \int_A dA$$

(L = length, A = cross-sectional area) or

$$U = \frac{1}{2EA} \int_L P^2 \, dx$$

If the axial load is constant over a length L, the strain energy absorbed by the member for this length is

$$U = \frac{P^2}{2EA} \int_L dx = \frac{P^2 L}{2EA} \tag{11}$$

Strain energy: torsional loading   In the case of the torsionally loaded member of Figure 9-4, the stress components σ_x, σ_y, σ_z, and τ_{yz} are all zero, so that the strain energy given by Equation (10) is

$$U = \tfrac{1}{2} \int_v (\tau_{xy}\gamma_{xy} + \tau_{xz}\gamma_{xz}) \, dv$$

Since the material is linearly elastic and obeys Hooke's law, the strain energy is

$$U = \frac{1}{2} \int_v \left(\tau_{xy}\frac{\tau_{xy}}{G} + \tau_{xz}\frac{\tau_{xy}}{G}\right) dv \quad \text{or} \quad U = \frac{1}{2G} \int_v (\tau_{xy}^2 + \tau_{xz}^2) \, dv$$

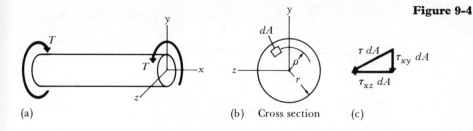

Figure 9-4

(a)   (b)   Cross section   (c)

From Figure 9-4(c), we see that the torsional shear stress may be expressed in terms of its rectangular components as follows:

$$(\tau\, dA)^2 = (\tau_{xz}\, dA)^2 + (\tau_{xy}\, dA)^2 \quad \text{or} \quad \tau^2 = \tau_{xz}^2 + \tau_{xy}^2$$

Thus the strain energy is

$$U = \frac{1}{2G} \int_v \tau^2\, dv \tag{12}$$

In the special case of a circular cross section,

$$U = \frac{1}{2G} \int_v \left(\frac{Tr}{J}\right)^2 dv$$

From Figure 9-5, we see that the differential volume may be expressed as

$$dv = dA\, dx \quad \text{or} \quad dv = \rho\, d\theta\, d\rho\, dx$$

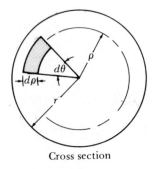

Figure 9-5

Cross section

Thus the strain energy becomes

$$U = \frac{1}{2G} \int_v \frac{T^2 \rho^2}{J^2} \rho\, d\theta\, d\rho\, dx$$

If the cross-sectional area is constant,

$$U = \frac{1}{2GJ^2} \int_L T^2\, dx \int_0^{2\pi} d\theta \int_0^r \rho^3\, d\rho$$

$$= \frac{\pi r^4}{4GJ^2} \int_L T^2\, dx = \frac{1}{2GJ} \int_L T^2\, dx$$

(Remember, $J = \pi r^4/2$.) If the torque T is constant over a length L, then the strain energy absorbed by the member of length L is

$$U = \frac{T^2}{2GJ}\int_L dx = \frac{T^2 L}{2GJ} \tag{13}$$

Strain energy: flexural loading The components of stress for the flexurally loaded member of Figure 9-6 are:

$$\sigma_x = \frac{My}{I}, \qquad \tau_{xy} = \frac{VQ}{It}, \qquad \sigma_y = \sigma_z = \tau_{xz} = \tau_{yz} = 0$$

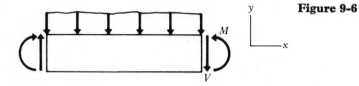

Figure 9-6

The strain energy, according to Equation (10), is

$$U = \frac{1}{2}\int_v (\sigma_x \epsilon_x + \tau_{xy}\gamma_{xy})\, dv = \frac{1}{2}\int_v \left[\sigma_x\left(\frac{\sigma_x}{E}\right) + \tau_{xy}\left(\frac{\tau_{xy}}{G}\right)\right] dv \tag{14}$$

or

$$U = \frac{1}{2}\int_v \left(\frac{\sigma_x^2}{E} + \frac{\tau_{xy}^2}{G}\right) dv$$

Let's look at the first term of Equation (14). For convenience, we refer to it as U_σ.

$$U_\sigma = \frac{1}{2E}\int_v \sigma_x^2\, dv = \frac{1}{2E}\int_v \left(\frac{My}{I}\right)^2 dv$$

Since the cross-sectional area is constant,

$$U_\sigma = \frac{1}{2EI^2}\int_v M^2 y^2\, dv$$

If we express the differential volume as

$$dv = dA\, dx$$

where A is the cross-sectional area, then

$$U_\sigma = \frac{1}{2EI^2}\int_v M^2 y^2\, dA\, dx = \frac{1}{2EI^2}\int_L M^2\, dx \int_A y^2\, dA$$

$$= \frac{1}{2EI^2}\int_L M^2\, dx\, (I) = \frac{1}{2EI}\int_L M^2\, dx \tag{15}$$

▶ **NOTE** The bending moment M in Equation (15) is usually not constant, but a function of x.

Now let's look at the second term of Equation (14). For convenience, we refer to it as U_τ.

$$U_\tau = \int_v \tau_{xy}^2 \, dv = \frac{1}{2G} \int_v \left(\frac{VQ}{It}\right)^2 dv$$

For a constant cross section, the moment of inertia I may be taken outside the integral. However, the quantities V, Q, and t are usually variables and must remain inside the integral.

$$U_\tau = \frac{1}{2GI^2} \int_v \frac{V^2 Q^2}{t^2} \, dv \tag{16}$$

Equation (16) is a general expression for all beams with constant cross sections. Let's investigate the special case of a rectangular cross section, as shown in Figure 9-7, where the quantity τ_{xy} of Equation (14) is the shear stress at an arbitrary distance y from the neutral axis. For the rectangular section, the statical moment Q is (see Chapter 4)

$$Q = \bar{y}A_F = \left[\frac{1}{2}\left(\frac{h}{2} + y\right)\right]\left[\left(\frac{h}{2} - y\right)t\right]$$

Figure 9-7

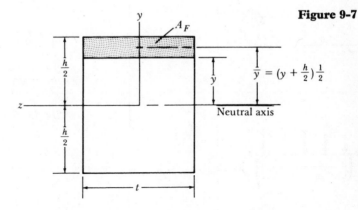

Thus Equation (16) becomes

$$U_\tau = \frac{1}{2GI^2} \int_v \frac{V^2}{t^2} \frac{t^2}{4} \left(\frac{h^4}{16} - \frac{h^2 y^2}{2} + y^4\right) dx \, dy \, dz$$

$$= \frac{1}{8GI^2} \int_0^t dz \int_{h/2}^{h/2} \left(\frac{h^4}{16} - \frac{h^2 y^2}{2} + y^4\right) dy \int_L V^2 \, dx$$

$$= \frac{1}{8GI^2}(t)\left(\frac{h^5}{30}\right)\int_L V^2 \, dx = \frac{3}{5GA}\int_L V^2 \, dx \tag{17}$$

We combine Equations (15) and (17) to obtain the expression for the total strain energy absorbed in the case of flexural loading of a beam with a uniform rectangular cross section.

$$U = \frac{1}{2EI} \int_L M^2 \, dx + \frac{3}{5GA} \int_L V^2 \, dx \tag{18}$$

▶ **NOTE** Unless the beam is very short, the second term of Equation (18) or Equation (14) is very small compared with the first term, and thus is neglected.

Summary of strain energy

Type of loading	1 * Strain energy	2 ** Stress	3 *** Strain energy
 Centric	$\dfrac{1}{2E} \int_v \sigma_x^2 \, dv$	$\sigma_x = \dfrac{P}{A}$	$\dfrac{1}{2AE} \int_L P^2(x) \, dx$ or $\dfrac{P^2 L}{2AE}$ if P is uniform over the length L
 Torsion	$\dfrac{1}{2G} \int_v \tau^2 \, dv$	$\tau = \dfrac{T\rho}{J}$	$\dfrac{1}{2JG} \int_L T^2(x) \, dx$ or $\dfrac{T^2 L}{2JG}$ if T is uniform over the length L
 Flexure	$\dfrac{1}{2E} \int_v \sigma_x^2 \, dv$ $+ \dfrac{1}{2E} \int_v \tau_{xy}^2 \, dv$	$\sigma_x = \dfrac{My}{I}$ $\tau_{xy} = \dfrac{VQ}{It}$	$\dfrac{1}{2EI} \int_L M^2(x) \, dx$ $+ \dfrac{3}{5GA} \int_L V^2(x) \, dx$ The second term is for rectangular cross sections only.

* The limitation on the general expressions for strain energy is that the material must be linearly elastic.
** Refer to previous chapters for the appropriate limitations on these stress equations.
*** The limitations on the expressions in column 3 are those for columns 1 and 2 plus any additional ones stated in column 3.

Castigliano's method for deflection

A structure subjected to external loads absorbs energy in the form of strain energy, as you learned in the previous pages of this section. The method of Castigliano uses strain energy in a mathematical sense to determine deflections of the structure.

We can most easily present this method by considering a specific configuration, such as a beam with several loads, as shown in Figure 9-8. Let's assume that P_1 and P_2 are applied to the beam simultaneously. The strain energy absorbed by the beam, which is equivalent to the work done by the external loads on the beam (Equation 2), may be expressed as

$$U = \tfrac{1}{2}P_1 y_1 + \tfrac{1}{2}P_2 y_2 \tag{19}$$

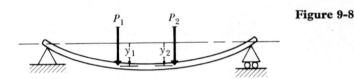

Figure 9-8

If we give the load P_1 a small increase of ΔP_1, the beam will deflect an additional amount, as shown in Figure 9-9. The strain energy absorbed by the beam from the application of the loads P_1 and P_2, plus the additional load ΔP_1, is

$$U + \Delta U = \tfrac{1}{2}P_1 y_1 + \tfrac{1}{2}P_2 y_2 + P_1\,\Delta y_1 \\ + P_2\,\Delta y_2 + \tfrac{1}{2}\Delta P_1\,\Delta y_1 \tag{20}$$

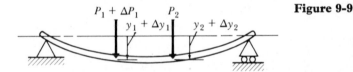

Figure 9-9

The last three terms, which correspond to ΔU, arise from applying the additional load ΔP_1 after loads P_1 and P_2 have already been applied. The first two terms correspond to the original work of P_1 and P_2 (Equation 19). The last term corresponds to the work done by the added force ΔP_1, the third and fourth terms to the additional work done by P_1 and P_2, respectively, in moving the beam the additional distances of Δy_1 and Δy_2.

Are you confused? Well, let's look at it from a different point of view.

The work done by the force P_1 (or the energy absorbed by the beam from the application of the force P_1) is the area under the P_1-versus-y curve; similarly for P_2 and ΔP_1. If we refer to these curves (Figure 9-10), and calculate the area under each, we see that the resulting total area corresponds to the strain energy given by Equation (20).

Now suppose we reverse the order of loading. First let's apply the load ΔP_1, as shown in Figure 9-11. The strain energy absorbed by the beam from ΔP_1 is

$$\Delta U = \tfrac{1}{2}P_1 y_1 \tag{21}$$

Figure 9-10

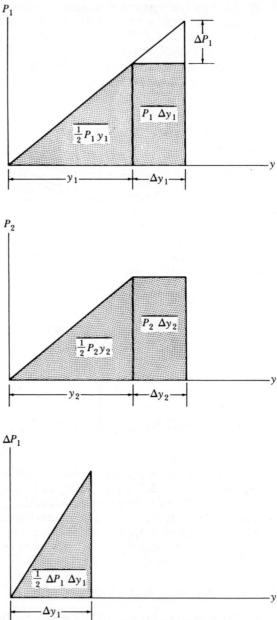

Then, if we apply the loads P_1 and P_2 simultaneously after P_1 has been applied, the beam will deflect to a final configuration as shown in Figure 9-9, and the total strain energy absorbed is

$$\Delta U + U = \tfrac{1}{2}\Delta P_1\,\Delta y_1 + \tfrac{1}{2}P_1 y_1 + \tfrac{1}{2}P_2 y_2 + \Delta P_1\,\Delta y_1 \qquad (22)$$

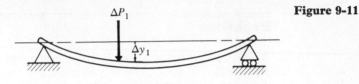

Figure 9-11

The final state of strain must be the same regardless of the order of loading. Thus Equations (22) and (20) are the same, and

$$P_1 \, \Delta y_1 + P_2 \, \Delta y_2 = \Delta P_1 y_1 \tag{23}$$

Now we have only a few mathematical manipulations left to complete the derivation of Castigliano's method. From Equations (19) and (20), we see that

$$\Delta U = P_1 \, \Delta y_1 + P_2 \, \Delta y_2 + \tfrac{1}{2}\Delta P_1 \, \Delta y_1 \tag{24}$$

If we substitute Equation (23) into (24), we have

$$\Delta U = y_1 \, \Delta P_1 + \tfrac{1}{2}\Delta P_1 \, \Delta y_1$$

or

$$\frac{\Delta U}{\Delta P_1} = y_1 + \frac{1}{2} \Delta y_1 \tag{25}$$

If we take the limit of Equation (25) as $\Delta P_1 \to 0$ (therefore also $\Delta y_1 \to 0$),

$$\frac{\partial U}{\partial P_1} = y_1 \tag{26}$$

It is necessary to use the partial derivative here if more loads than one are present. For the general case in which a structure has many loads, Equation (26) becomes

$$\frac{\partial U}{\partial P_i} = y_i \tag{27}$$

where y_i is the deflection of the point on the structure at which the load P_i is applied and in the direction of P_i. Similarly, we can show the method to be valid for applied moments:

$$\frac{\partial U}{\partial M_i} = \theta_i \tag{28}$$

where θ_i is the rotation in radians of the point on the structure at which the moment M_i is applied and in the direction of M_i.

Turn to the Study Guide, where you will find more explanation and some examples of the use of Equations (27) and (28). ■ STOP

Problems

9-1.1 A long slender, straight beam is homogeneous and isotropic, obeys Hooke's law, and has a constant cross section. Determine the lateral deflection at the 6000-lb load. The beam is S 5 × 10.

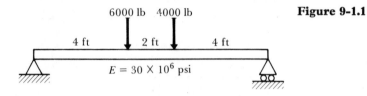

Figure 9-1.1

9-1.2 For the pin-connected truss shown, determine the displacement of point B.

Area	Member
2 in^2	AB
4 in^2	CB
$E = 30 \times 10^6$ psi	

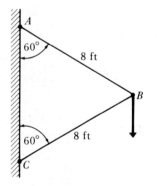

Figure 9-1.2

9-1.3 A straight beam is homogeneous and isotropic, obeys Hooke's law, and has a constant cross section

a) Suppose that the beam is long and slender. Determine the lateral deflection and slope at the free end.

b) Suppose that the beam is short, with a large rectangular cross section. Determine the lateral deflection and slope at the free end.

c) Determine the ratio of length to cross-sectional radius of gyration for which the deflection of part (a) deviates by 10% from the deflection of part (b).

9-1.4 Determine the rotation of the torque T_0 at the free end of the frame of uniform circular cross section. E, I, J, and G are constant.

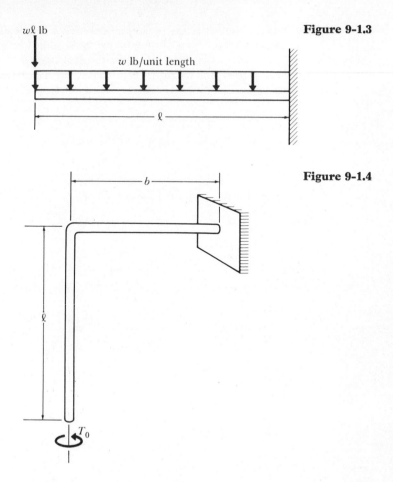

Figure 9-1.3

$w\ell$ lb

w lb/unit length

ℓ

Figure 9-1.4

b

ℓ

T_0

9-1.5 Determine the deflection of point A of the frame. The three sides of the frame are considered long and slender.

Figure 9-1.5

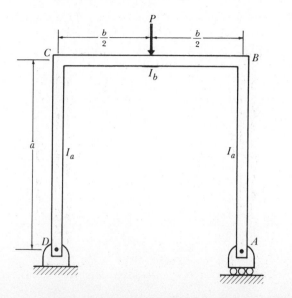

P

$\dfrac{b}{2}$ $\dfrac{b}{2}$

C B

I_b

a

I_a I_a

D A

9-1.6 Each member of the pin-connected truss is 8 ft long and has a modulus of elasticity of 30,000,000 psi. Determine the horizontal and vertical components of the deflection of point C.

Cross-sectional area	Member
2 in²	AB
10	DB
5	DC
5	BC
4	AD

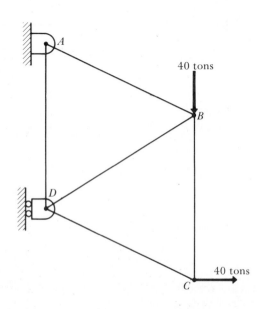

Figure 9-1.6

9-1.7 The slender uniform circular bar of radius r is in the shape of a semicircle with a radius R. It is fixed at one end and is subjected to a torque T_0 at the free end. Determine the angle through which the torque T_0 twists.

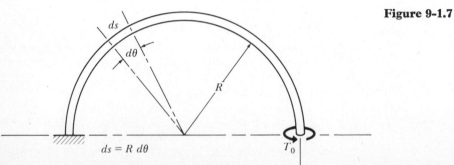

Figure 9-1.7

9-2.1 The long steel beam is fixed at A and supported at B by an aluminum tie rod. Determine the force in the rod and the reactions on the beam at A.

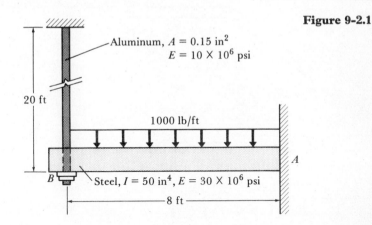

Figure 9-2.1

9-2.2 The beam of Problem 9-2.1 is also supported by a second aluminum rod at midspan. What are the force in each rod and the reactions at A?

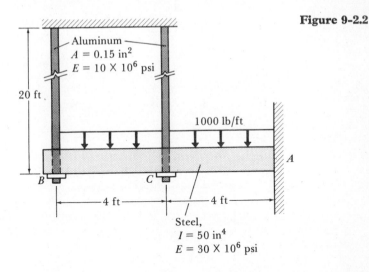

Figure 9-2.2

9-2.3 A long slender L-shaped steel rod, whose cross-sectional diameter is 2 in., is built in at one end to a rigid wall and simply supported at the other end. The shaft has a 90° bend in a horizontal plane. Determine the reactions of the shaft at each end.

9-2.4 The reaction at C is zero before the load w is applied. After the load has been applied, what are the reactions at E, C, and D? Determine the deflection at C. Neglect the strain energy of flexural shear loading.

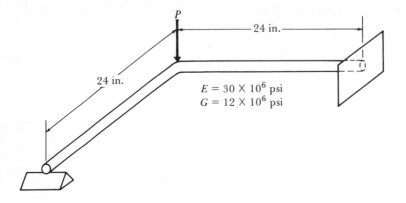

Figure 9-2.3

P

— 24 in. —

24 in.

$E = 30 \times 10^6$ psi
$G = 12 \times 10^6$ psi

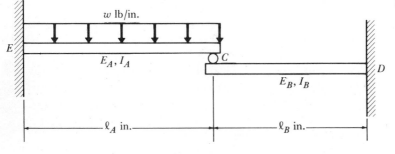

Figure 9-2.4

w lb/in.

E

E_A, I_A

C

E_B, I_B

D

— ℓ_A in. —

— ℓ_B in. —

9-2.5 A given frame is pinned at A and B. Neglect the strain energy of centric loading and flexural shear loading and determine the reactions at A and B. (*Hint:* Make use of symmetry.)

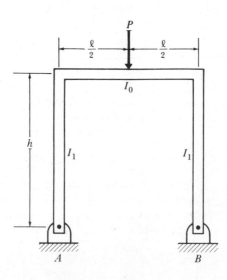

Figure 9-2.5

P

$\frac{\ell}{2}$

$\frac{\ell}{2}$

I_0

h

I_1

I_1

A

B

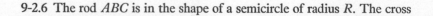

9-2.6 The rod ABC is in the shape of a semicircle of radius R. The cross section is circular with a radius r, which is very small compared with R. Determine the reactions at A and C, where the rod is fixed. (*Hint:* Make use of symmetry.)

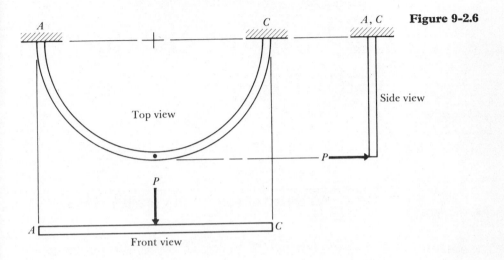

Figure 9-2.6

Chapter ten

Stability of elastic columns

R10-1 Stability of centrically loaded elastic columns

The stability concept

We can illustrate the concept of stable and unstable equilibrium by examining a smooth sphere combined with a smooth bowl, as shown in Figure 10-1.

Weight of sphere

Figure 10-1

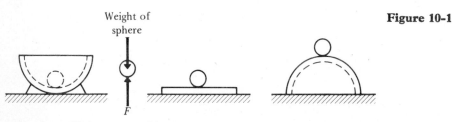

F

(a) Stable equilibrium   (b) Neutral equilibrium   (c) Unstable equilibrium

The sphere shown in Figure 10-1(a) is in equilibrium, since the reacting force F equals the weight of the sphere. This sphere is said to be in a condition of *stable equilibrium*, since, if the sphere is given a small displacement in any direction, it will roll back to its original equilibrium position. As the curvature of the bowl is reduced, the degree of stability is also reduced. In the limit when the bowl becomes a flat surface (Figure 10-1b), the system will be in a condition or configuration of *neutral equilibrium*. This means that, if the ball is displaced, it will not return to its original position, nor will it move to a new position. The ball will simply remain in its displaced position. Finally, as the curvature of the bowl is reversed, the system assumes an *unstable equilibrium* configuration, as shown in Figure 10-1(c). In this configuration, any displacement of the ball will result in its continued displacement from its original equilibrium configuration. As engineering designers, we will want to avoid structural systems that are unstable, as in Figure 10-1(c).

In this section, our primary interest is to determine the loads at which elastic columns (such as a yardstick) become unstable. We can approximate the stability behavior of an elastic column by studying a rigid bar with a torsional spring attached at one end, and loaded as shown in Figure 10-2. We shall examine the case of a rigid bar with a torsional spring before we consider elastic columns, since both types of problems exhibit similar behaviors and the analysis of the rigid bar–spring is mathematically and intuitively more straightforward than that of the elastic column. Consequently, we shall be able to acquaint ourselves with the concept of the general stability problem before dealing with the mathematical specifics of an elastic column.

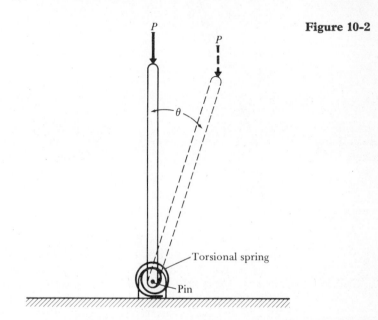

Figure 10-2

Let's assume that the circular spring illustrated in Figure 10-2 has a spring constant of k (in.-lb/rad), and that it is equally resistant to rotation in either the clockwise or the counterclockwise directions. Recall that systems are said to be unstable if they collapse when subjected to a very small displacement. Thus let's assume that you have grasped the rigid bar in your hand, and have displaced the bar through a very small angle θ ($\sin \theta \approx \theta$) to the position shown in Figure 10-3.

We want to determine the load at which this system changes from stable to unstable equilibrium, that is, the point of neutral equilibrium. Thus we want to determine the load at which the bar will remain in equilibrium after it is released in the slightly deflected position. We can determine the load P which corresponds to the condition of neutral equilibrium for the bar by requiring that the summation of moments be equal to zero for the loads shown in Figure 10-3.

$$\sum M_A = 0$$

$$-k\theta + P\ell\theta = 0 \qquad \text{(since } \sin \theta \approx \theta \text{ for small } \theta\text{)}$$

or

$$P = P_{cr} = k/\ell \tag{1}$$

where the load corresponding to neutral equilibrium (P_{cr}) is called the *critical* or *buckling load*.

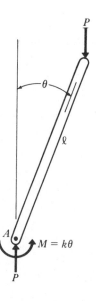

Figure 10-3

▶ **NOTE** Let's assume that we have grasped the bar, on which the load P acts, and rotated it into a position corresponding to some small angle θ. If P is smaller than k/ℓ, then the bar will spring back into its initial vertical equilibrium position when it is released. Thus, by our previous definitions, the bar and spring are in a state of stable equilibrium. However, if P is larger than k/ℓ, the bar will continue to rotate (collapse) when it is released. Under these conditions, the system is said to be unstable. Thus the maximum allowable load that the bar–spring system can resist without collapse is $P_{cr} = k/\ell$.

If the bar is deflected far enough and θ becomes so large that $\sin \theta \approx \theta$ no longer holds, then the linearized model given by Equation (1) becomes inaccurate. Consequently, if we are to consider large angles of rotation, the equation for moment equilibrium becomes

$$\Sigma M_A = 0$$

$$-k\theta + P\ell \sin \theta = 0$$

or

$$P = \frac{k}{\ell}\left(\frac{\theta}{\sin \theta}\right) \tag{2}$$

Equation (2) is plotted in Figure 10-4.

Figure 10-4

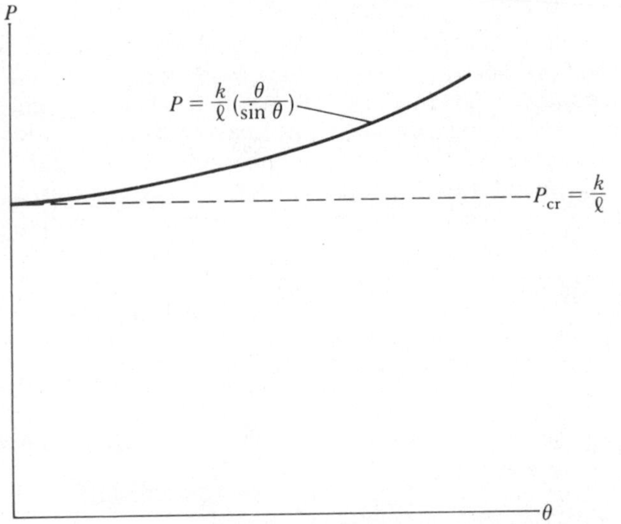

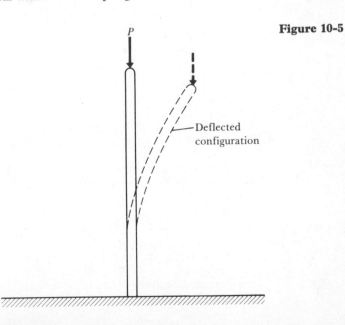

$$P = \frac{k}{\ell}\left(\frac{\theta}{\sin \theta}\right)$$

$$P_{cr} = \frac{k}{\ell}$$

▶ **NOTE** From Figure 10-4, we observe that the bar and spring will carry some additional load beyond the buckling load P_{cr}. However, the deflections become relatively large with only slight increases in load above P_{cr}. Such large deflections are usually unacceptable, and for all practical purposes, P_{cr} is the maximum allowable load.

Elastic columns   The behavior of the perfectly straight centrically loaded elastic column shown in Figure 10-5 is highly analogous to the behavior of the rigid bar with a circular spring.

Figure 10-5

Deflected configuration

▶ **NOTE** A column is a structural or machine member that is axially loaded in compression and whose length is much greater than its cross-sectional dimensions.

The resistance to lateral deflection modeled by the spring in the bar–spring problem is analogous to the bending resistance of an elastic column. We can determine the buckling load of an elastic column by visualizing the column in a slightly deflected configuration, and then requiring that the column be in equilibrium (neutral equilibrium), just as we did with the bar–spring system. Any loads that correspond to the configuration of deflected (neutral) equilibrium are buckling loads.

The buckling behavior of the fixed–free column shown in Figure 10-6(a) is precisely the same as the buckling behavior of a column with twice the length and pinned ends, as shown in Figure 10-6(b).

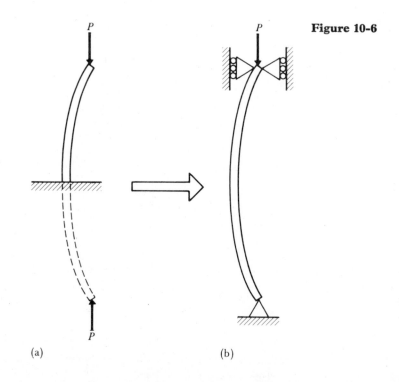

Figure 10-6

(a)   (b)

The derivation of buckling loads for the columns shown in Figure 10-6(a) and 10-6(b) follow the same general pattern, but the results are somewhat more easily interpreted as corresponding to the column with both ends pinned (Figure 10-6b). Thus let's derive an expression for the pinned column of Figure 10-6(b).

For our first step, let's require that the column be in equilibrium in the slightly deflected configuration shown in Figure 10-7. The bending moment M and the axial load P are the internal reactions acting on the cross section.

Figure 10-7

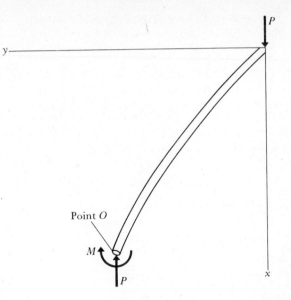

▶ **NOTE** We assume the bending moment in Figure 10-7 to be in the positive direction, as defined in Chapter 5, thus preserving consistency.

If the column segment in Figure 10-7 is in equilibrium then we can write $\sum M_0 = 0$. Thus

$$M + Py = 0$$

Since

$$M = EI \frac{d^2y}{dx^2}$$

we have

$$\frac{d^2y}{dx^2} + \frac{P}{EI} y = 0 \tag{3}$$

▶ **NOTE a)** As discussed in Chapter 5, the problem is limited to small deflections (linearized) by the assumption that

$$M = EI \frac{d^2y/dx^2}{[1 + (dy/dx)^2]^{3/2}} \approx EI \frac{d^2y}{dx^2}$$

b) The assumptions made in the column theory are the same as the assumptions made in developing the beam-deflection theory in Chapter 5.

Equation (3) is a linear differential equation with constant coefficients, and has a solution of the form

$$y = Ae^{\lambda x} \tag{4}$$

where A and λ are constants. Substitution of Equation (4) into Equation (3) yields

$$A\lambda^2 e^{\lambda x} + \frac{P}{EI}\, Ae^{\lambda x} = 0$$

Thus

$$\lambda = \pm \sqrt{\frac{-P}{EI}} = \pm\, i\, \sqrt{\frac{P}{EI}}$$

where $\sqrt{-1} = i$. Consequently, the solution to Equation (3) is

$$y = A_1 e^{-i\sqrt{P/EI}\,x} + A_2 e^{i\sqrt{P/EI}\,x} \tag{5}$$

If we use the trigonometric identities

$$\sin\sqrt{\frac{P}{EI}}\, x = \frac{e^{i\sqrt{P/EI}\,x} - e^{-i\sqrt{P/EI}\,x}}{2i}$$

and

$$\cos\sqrt{\frac{P}{EI}}\, x = \frac{e^{i\sqrt{P/EI}\,x} + e^{-i\sqrt{P/EI}\,x}}{2}$$

we may write Equation (5) in the form

$$y = a_1 \sin\sqrt{\frac{P}{EI}}\, x + a_2 \cos\sqrt{\frac{P}{EI}}\, x \tag{6}$$

where a_1 and a_2 are constants.

$$a_1 = A_1 + A_2, \qquad a_2 = (A_1 - A_2)i$$

Our next step is to determine the constants a_1 and a_2 by requiring that Equation (6) satisfy the constraints on the column. A column with both ends pinned is subjected to the following two constraints (boundary conditions):

1 $y = 0$ at $x = 0$
2 $y = 0$ at $x = \ell$

Substitution of the first condition ($y = 0$ at $x = 0$) into Equation (6) yields

$$0 = a_1(0) + a_2(1) \quad \text{or} \quad a_2 = 0$$

Then substitution of the second condition ($y = 0$ at $x = \ell$) and $a_2 = 0$ into Equation (6) results in

$$0 = a_1 \sin\sqrt{P/EI}\;\ell \tag{7}$$

The trivial solution to Equation (7), $a_1 = 0$, gives us no usable information. Thus we must have

$$\sin \sqrt{P/EI}\, \ell = 0$$

or

$$\ell \sqrt{P/EI} = n\pi \tag{8}$$

where $n = 1, 2, 3, 4, \ldots$. Solving Equation (8) gives us the possible buckling loads for a column with both ends pinned:

$$P = P_{\mathrm{cr}} = \frac{n^2 \pi^2 EI}{\ell^2}, \qquad n = 1, 2, 3, \ldots \tag{9}$$

If we examine Equation (9), we will see that there are an infinite number of buckling loads (one corresponding to each value of n). The column will first buckle and therefore fail when the compressive load reaches the lowest (first) buckling load. Thus, for a column with pinned ends, the lowest buckling load corresponds to $n = 1$, and we have

$$P_{\mathrm{cr}} = \frac{\pi^2 EI}{\ell^2} \tag{10}$$

or

$$\frac{P_{\mathrm{cr}}}{A} = \frac{\pi^2 E}{(\ell/r)^2} \tag{11}$$

where A is the cross-sectional area of the column and r is the smallest radius of gyration ($I = Ar^2$) corresponding to the cross-sectional area.

▶ **NOTE** Equation (11) or (10) is frequently referred to as *Euler's equation*, in honor of the Swiss mathematician Leonard Euler, who first developed this equation in 1757. The ratio ℓ/r is called the *slenderness ratio*.

Corresponding to $n = 1$, Equation (6) becomes

$$y = a_1 \sin \frac{\pi x}{\ell} \tag{12}$$

Again, we observe the analogy between the elastic column and the bar–spring problem. Equations (10) and (12) were derived from a linearized solution, because of the assumption of small deflections. Consequently, Equation (10) is independent of deflection, just as Equation (1) is in the bar–spring problem. We can also see that a_1 is arbitrary in Equation (12), which implies that the deflections could go to infinity without any increase in compressive load. If we account for the nonlinearity of large deflections (which is a subject beyond the scope of this text), we shall obtain a P-versus-$y_{\max}$ relationship very similar to the one shown in Figure 10-4 for the bar–spring system. Thus it is important to note that the lateral deflections of an elastic column are very large, corresponding to very small increases in load above P_{cr}, and for most practical purposes, the buckling load is the failure load. In fact, lateral deflections due to buckling are frequently so large that the failure is often catastrophic when the buckling load is exceeded only slightly. As a result, most structures are designed to operate with loads far below any buckling loads.

Buckling in modes corresponding to $n = 2, 3, 4, \ldots$ will occur only if additional constraints are added to the column. As an example, Figure 10-8 illustrates buckling modes corresponding to $n = 2$ and $n = 3$.

Figure 10-8

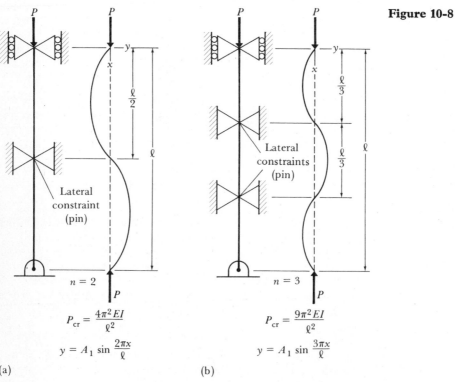

(a)   (b)

The additional lateral constraints generally consist of relatively light structural members compared to the column itself. In cases in which lateral constraints are structurally feasible, they are a good technique with which to achieve additional strength.

Similar derivations can be developed for columns with their ends constrained in ways other than having both ends pinned. In each case, the result can be reduced to the form

$$P_{cr} = C \frac{\pi^2 EI}{\ell^2} = C \frac{\pi^2 EA}{(\ell/r)^2} \tag{13}$$

where C is a constant dependent on the end constraints. Table 10-1 gives values of the constant C, corresponding to several sets of end constraints.

Column end constraints	C	ℓ_e	
Both ends pinned	1	ℓ	
One end clamped (fixed) One end free	0.25	2ℓ	
One end clamped (fixed) One end pinned	2.05	0.7ℓ	
Both ends clamped (fixed)	4	$\dfrac{\ell}{2}$	

▶ **NOTE** Often in the literature on mechanics of materials, the factor C is combined with the length to define an effective length ℓ_e, where $\ell_e = \ell/\sqrt{C}$. As a result, Equation (13) is often written in the form

$$P_{cr} = \frac{\pi^2 EA}{(\ell_e/r)^2} \tag{14}$$

Note that the effective length ℓ_e is the distance between two points of zero bending moment.

As an example, a column with both ends fixed would have an effective length of

$$\ell_e = \frac{\ell}{\sqrt{4}} = \frac{\ell}{2}$$

Frequently, in actual practice, it is very difficult to determine the proper way of fixing the end (and thus the effective length) of a column. With the exception of the case of a column with one free end and one fixed end, it is safe to assume that the ends must be pinned. Since stability failures are frequently catastrophic, it is rather common practice, for the sake of safety, to be conservative in stability analyses.

Equation (13), which is the Euler equation, is derived for columns with constant cross section that are loaded centrically. That is, the load application is coincident with the longitudinal centroidal axis. In reality, we might expect most columns to be loaded with at least some eccentricity (often accidental). However, experiments with many columns show that the effects of eccentric loading are not important for long, slender columns, and consequently, the Euler equation accurately predicts buckling loads for long, slender columns. In this section, we shall define the *long slender range* for columns to include steel columns for which $\ell_e/r > 140$, and aluminum or wood columns for which $\ell_e/r > 80$ (see Figure 10-9).

▶ **NOTE** These limits for ℓ_e/r are approximate, and are given only as a general lower bound of the long slender range. For many materials, empirical equations have been developed that define the specific value of ℓ_e/r corresponding to the lower bound of the long-slender-column range.

When ℓ_e/r is very small, the column behaves as a compression block, and the failure load is determined by the compressive strength of the material ($P = \sigma_{YP}A$). The value of ℓ_e/r corresponding to the upper bound of the compression-block range (see Figure 10-9) is determined as follows:

$$\frac{P}{A} = \sigma_{YP} = \frac{\pi^2 E}{(\ell_e/r)^2} \quad \text{or} \quad \ell_e/r = \sqrt{\frac{\pi^2 E}{\sigma_{YP}}}$$

Figure 10-9

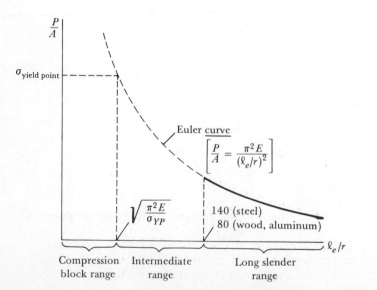

It is probably not too surprising to find that the region between the long slender range and the compression-block range is called the intermediate range (see Figure 10-9). Equations suitable for predicting the buckling behavior of columns in the intermediate range are discussed in Section R10-2 and R10-3.

Turn to Section SG10-1 of the Study Guide. ■ **STOP**

R10-2 Stability of eccentrically loaded elastic columns with pinned ends

In the previous section, we discussed the special case of a centrically loaded column. We also noted that in any actual centric loading of a column, the load is most probably applied with at least a small eccentricity, either accidental or intentional. Experiments with columns show that Euler's equation accurately predicts buckling loads for long slender columns, but where ℓ_e/r is in the intermediate range, it is necessary to include the effects of eccentric loading. Thus in this section, we shall consider eccentric loading.

Let's examine an elastic column on which the compression load acts at a known distance e from the center line of the unloaded column, as shown in Figure 10-10(a). (*Note:* For small deflections, the distance e in Figure 10-10 is assumed to remain invariant.)

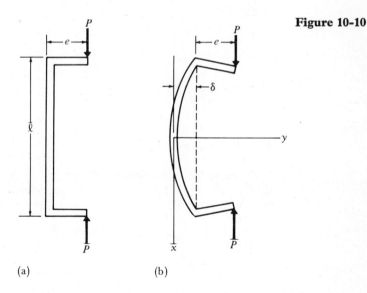

Figure 10-10

(a)   (b)

▶ **NOTE** When a column has an eccentric load, even the smallest load P will produce bending moments in the column, and then lateral deflections. Consequently, the assumption of no lateral deflections until the buckling load is reached, at which point there are instantaneous large deflections, is not valid for a column with an eccentric load. We shall find that the buckling load $P_{cr} = C\pi^2 EI/\ell^2$ remains as an upper limit, but that the large lateral deflections sometimes occur at load levels well below this limit.

To determine the load–deflection relationship for an eccentrically loaded column, we begin by obtaining the expression for bending moment from the free-body diagram shown in Figure 10-11. We shall again limit our derivation to small deflections. The deflections shown in Figure 10-11 are highly exaggerated for the sake of illustration.

Figure 10-11

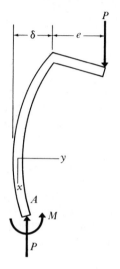

$$\sum M_A = 0$$

$$M - P(e + \delta - y) = 0$$

or

$$M = P(e + \delta - y)$$

In the beam theory developed in Chapter 5, we obtained the equation $EI(d^2y/dx^2) = M$. Thus we can write the differential equation for the elastic curve as

$$EI\frac{d^2y}{dx^2} = M = P(e + \delta - y)$$

or

$$\frac{d^2y}{dx^2} + \frac{P}{EI}y = \frac{P}{EI}(e + \delta) \tag{15}$$

▶ **NOTE** Since we used the equation $EI(d^2y/dx^2) = M$, the limitations specified for linearized beam theory also apply to column theory.

The solution to Equation (15) can be divided into homogeneous and particular parts, $y = y_h + y_p$, where y_h is the solution to

$$\frac{d^2 y_h}{dx^2} + \frac{P}{EI} y_h = 0 \tag{16a}$$

and y_p is the particular solution to

$$\frac{d^2 y_p}{dx^2} + \frac{P}{EI} y_p = \frac{P}{EI}(e + \delta) \tag{16b}$$

We should note that Equation (16a) is of the same form as Equation (3) in Section R10-1. Consequently, based on our work in Section R10-1, we can write the solution

$$y_h = a_1 \sin \sqrt{\frac{P}{EI}} x + a_2 \cos \sqrt{\frac{P}{EI}} x \tag{17}$$

By inspection of Equation (16b), we can determine the particular solution

$$y_p = e + \delta \tag{18}$$

▶ **NOTE** We assume that you have some previous knowledge of differential equations. If, however, you are confused by any or all of the steps leading to the solution to Equation (15), you should refer to a text on differential equations.

We obtain the general solution to Equation (15) by combining Equations (17) and (18):

$$y = a_1 \sin \sqrt{\frac{P}{EI}} x + a_2 \cos \sqrt{\frac{P}{EI}} x + e + \delta \tag{19}$$

Then we determine the coefficients a_1 and a_2 by using the conditions of zero deflection ($y = 0$) and zero slope ($dy/dx = 0$) at the origin ($x = 0$). Remember that the origin of the xy coordinate system is at the center of the column (Figures 10-10 and 10-11). Thus substitution of $x = 0$ and $y = 0$ into Equation (19) yields

$$0 = a_2 + e + \delta \qquad \text{or} \qquad a_2 = -e - \delta$$

Substitution of $dy/dx = 0$ at $x = 0$ yields

$$\frac{dy}{dx}\bigg|_{x=0} = \left[a_1 \sqrt{\frac{P}{EI}} \cos \sqrt{\frac{P}{EI}} x - a_2 \sqrt{\frac{P}{EI}} \sin \sqrt{\frac{P}{EI}} x \right]_{x=0} = 0$$

or

$$a_1 \sqrt{\frac{P}{EI}} = 0$$

Hence $a_1 = 0$. Substituting $a_1 = 0$ and $a_2 = -(e + \delta)$ into Equation (19) gives

$$y = (e + \delta)\left(1 - \cos\sqrt{\frac{P}{EI}}\,x\right) \tag{20}$$

The maximum deflection is $y = \delta$ at $x = \ell/2$. Thus

$$\delta = e + \delta - e\cos\frac{\ell}{2}\sqrt{\frac{P}{EI}} - \delta\cos\frac{\ell}{2}\sqrt{\frac{P}{EI}}$$

or

$$\delta = e\left(\frac{1 - \cos\frac{1}{2}\ell\sqrt{P/EI}}{\cos\frac{1}{2}\ell\sqrt{P/EI}}\right)$$

$$\delta = e\left(\sec\frac{\ell}{2}\sqrt{\frac{P}{EI}} - 1\right) \tag{21}$$

We can obtain the combined compressive stress in the column by superposition of the uniform compressive stress and the bending stress. Thus we write

$$\sigma = -\frac{P}{A} - \frac{Mc}{I} = -\frac{P}{A} - \frac{P(e + \delta)c}{Ar^2} \tag{22}$$

where c is the distance from the neutral axis to the outer fiber and r is the smallest radius of gyration of the column cross section ($I = Ar^2$). From Equation (21), we obtain

$$e + \delta = e\sec\frac{\ell}{2}\sqrt{\frac{P}{EI}} \tag{23}$$

Then substitution of Equation (23) into Equation (22) gives

$$\sigma = -\frac{P}{A}\left(1 + \frac{ec}{r^2}\sec\frac{\ell}{2}\sqrt{\frac{P}{EI}}\right) \tag{24}$$

or

$$\frac{P}{A} = \frac{-\sigma}{[1 + (ec/r^2)\sec\frac{1}{2}\ell\sqrt{P/EI}]} \tag{25a}$$

or

$$\frac{P}{A} = \frac{-\sigma}{[1 + (ec/r^2)\sec(\ell/2r)\sqrt{P/EA}]} \tag{25b}$$

Either of Equations (25) is known as the secant formula, and the quantity ec/r^2 is called the *eccentricity ratio*.

▶ **NOTE**   Equations (24), (25a), and (25b) can be applied to columns whose ends are not pinned, if ℓ_e is substituted for ℓ.

● **CAUTION** The ratio P/A appears on both sides of the equals sign in Equation (25b). Thus the solution for P/A must be determined by trial and error (iteration). In addition, you should note that P/A is not linearly related to σ. Consequently, if σ is changed by some proportion, the corresponding load will not be changed by the same proportion. Thus, *if we want to apply a factor of safety, we should apply it to the load and not the maximum stress.*

Equation (25b) can be most easily understood when it is presented graphically. Figure 10-12 contains a set of graphs corresponding to Equation (25b) for steel with $E = 30 \times 10^6$ psi and $\sigma = \sigma_{YP} = 40,000$ psi.

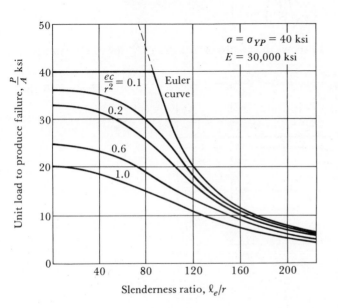

Figure 10-12

You can more clearly understand the behavior of an eccentrically loaded column (as described by the secant formula) if you observe that the buckling load for a column with centric loading is also the maximum load for the same column if the loading is eccentric. To demonstrate this, we must examine Equation (21), and observe that the deflection δ is infinite when $(\ell/2)\sqrt{P/EI}$ equals $\pi/2, 3\pi/2, 5\pi/2, \ldots$, since at these values of $(\ell/2)\sqrt{P/EI}$, sec θ equals infinity.

Consequently, the maximum deflection approaches infinity (independent of the value of e) when $(\ell/2)\sqrt{P/EI}$ approaches $\pi/2, 3\pi/2, 5\pi/2, \ldots$.

We can determine the maximum load corresponding to a column that buckles into the shape of the first mode, since we have

$$\frac{\ell}{2}\sqrt{\frac{P}{EI}} = \frac{\pi}{2}$$

from which

$$P_{cr} = \frac{\pi^2 EI}{\ell^2} \qquad (26)$$

Thus we see that, as the column deflects toward infinity, the corresponding load asymptotically approaches the Euler buckling load as an upper limit.

Substitution of Equation (26) into Equation (21) yields the following expression for the maximum lateral deflection of the column:

$$\delta = e \left[\sec \left(\frac{\pi}{2} \sqrt{\frac{P}{P_{cr}}} \right) - 1 \right] \qquad (27)$$

Figure 10-13 shows the graphs of Equation (27) for various values of e.

Figure 10-13

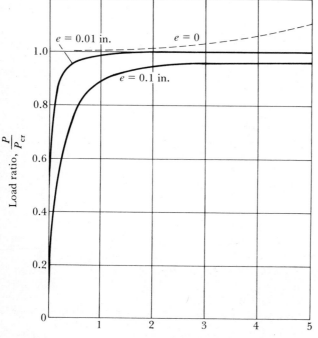

Lateral deflection δ, in.

Note that the lateral deflection approaches infinity as P approaches P_{cr}, regardless of the value of the eccentricity e. However, as e becomes small, the deflection stays very small until the load approaches the buckling load, at which time the deflections increase very rapidly for small increases in load. Thus columns with loads that are applied with only slight eccentricities exhibit a behavior very closely resembling the behavior of a column with a centric load. As a result, the stability failure of an actual column may still occur as a rather sudden (often catastrophic) event even with eccentric loading. Sudden application of loads above the buckling load will lead to an immediate and total collapse of the column.

Remember, the equations developed in this section are valid for small deflections. Thus our discussion of large deflections is somewhat suspect. However, the additional load-carrying capacity of a column loaded beyond the buckling load is very small (see dashed line in Figure 10-13). Note the similarity between the curve for $e = 0$ in Figure 10-13 and that in Figure 10-4. Thus the load–deflection curve that can be obtained from a large-deflection analysis is generally only of academic interest, since the accompanying deflections are too large to be acceptable for most real columns.

Summary

The following important relationships apply to an *eccentrically* loaded column with pinned ends:

$$\frac{P}{A} = \frac{-\sigma}{1 + (ec/r^2)\sec(\tfrac{1}{2}\ell\sqrt{P/EI})} = \frac{-\sigma}{1 + (ec/r^2)\sec(\ell/2r)\sqrt{P/EA}}$$

$$\delta = e\left[\sec\left(\frac{\pi}{2}\sqrt{\frac{P}{P_{cr}}}\right) - 1\right]$$

where

P = axial load
A = cross-sectional area
$\sigma = -P/A - Mc/I$ = combined compressive stress
e = eccentricity of the load P
c = distance from neutral axis to outer fiber
r = least radius of gyration of cross section
$I = Ar^2$ = least moment of inertia of cross section
E = modulus of elasticity
$P_{cr} = \pi^2 EI/\ell^2$ = buckling load (critical load)
δ = maximum lateral deflection of the column

Turn to Section SG10-2 of the Study Guide. ■ **STOP**

R10-3   Stability of centrically loaded elastic columns, empirical formulas

Engineers often find it very useful to develop equations based on experiments or observations. Such equations are called *empirical equations*.

We can visualize the need for empirical stability formulas by first summarizing the information that we developed in our theoretical derivations. Figure 10-12 illustrates the column equations that we developed. The Euler curve and the horizontal yield-stress line are developed for centrically loaded columns, and represent the maximum possible values of compressive column loads. Eccentric loading or structural imperfection may lead to large stresses and/or large lateral deflections at loads lower than the buckling loads. As a result, we must be rather cautious about the use of the equations that we developed for centrically loaded columns. Since eccentricities and/or structural imperfections always exist, the theoretical values will indicate stresses or deflections lower than those which actually exist.

We attempted to account for loading eccentricities by using the secant formula. However, if we are to use the secant formula, we must know the value of the eccentricity e, which is frequently accidental. Indeed, we often have no knowledge of the eccentricity or any structural imperfections. Therefore, if we are to account for eccentricities and structural imperfections, and if we are to evaluate the reliability of our buckling equations for centrically loaded columns, it is necessary for us to study the stability of elastic columns from an empirical viewpoint.

Numerous experiments with columns indicate that, in the case of very long slender columns, the effects of loading eccentricities and structural imperfections are minimal. Thus, for long slender columns, the Euler equation will predict the maximum allowable load with sufficient accuracy.

At the lower end of the slenderness-ratio scale, experiments show that the columns behave much as compression blocks, and failure is determined by the yield stress. In cases of such short columns, the deflection due to bending is relatively small. The stres scan be computed by means of the equation

$$\sigma = -\frac{P}{A} - \frac{My}{I} \quad \text{or} \quad \sigma = -\frac{P}{A} - \frac{Pey}{I}$$

It seems reasonable that we should use the secant equation, or an appropriate analogue of the secant equation, to describe the behavior of columns in the intermediate range between short and slender. Many empirical equations have been developed to describe the behavior of columns in the intermediate range. The most frequently used empirical column equations that predict buckling loads have the parabolic form

$$\frac{P}{A} = -\sigma_0 \left[1 - C_1 \left(\frac{\ell_e}{r} \right)^2 \right] \tag{28}$$

or the linear form

$$\frac{P}{A} = -\sigma_0 \left[1 - C_2 \left(\frac{\ell_e}{r} \right) \right] \tag{29}$$

where $\ell_e = \ell / \sqrt{C}$ is the effective column length. It can be shown that Equation (28) very closely matches the secant equation [Equation (25b) in Section 10-2] in the intermediate column range. As a consequence, Equation (28) should be appropriate for determining buckling loads.

Let's examine the empirical Equations (28) and (29) for two common structural metals, steel and aluminum.

Empirical column formula for steel

The American Institute of Steel Construction (*AISC Steel Construction Manual*, New York: AISC Inc., 1963) recommends that the upper limit of the intermediate-column range be the slenderness ratio that corresponds to $\sigma_{YP}/2$ on the Euler curve. Thus, if we rewrite the Euler equation in the form

$$\frac{\ell_e}{r} = \sqrt{\frac{\pi^2 EA}{P_{cr}}} = \sqrt{\frac{\pi^2 E}{\sigma}}$$

and if we let $\sigma = \sigma_{YP}/2$, we can derive the following equation for the slenderness ratio at the upper limit of the intermediate-column range:

$$\left(\frac{\ell_e}{r}\right)_{lim} = \sqrt{\frac{2\pi^2 E}{\sigma_{YP}}} \tag{30}$$

The AISC also recommends that a factor of safety (FS) of 1.92 be used for slender steel columns. Consequently, for any steel column for which $\ell_e/r \geq (\ell_e/r)_{lim}$, we use the following form of Euler's equation to compute the maximum allowable load:

$$\sigma_{cr} = -\frac{P_{cr}}{A} = -\frac{\pi^2 E}{1.92(\ell_e/r)^2} \tag{31}$$

When a steel column has a slenderness ratio such that $\ell_e/r < (\ell_e/r)_{lim}$, the AISC recommends the use of Equation (28) with $C_1 = \frac{1}{2}[(\ell_e/r)_{lim}]^2$ and $\sigma_0 = \sigma_{YP}$. If we also include a factor of safety, then Equation (28) assumes the form

$$\sigma_{cr} = -\frac{P}{A} = -\frac{\sigma_{YP}\{1 - \frac{1}{2}(\ell_e/r)^2[(\ell_e/r)_{lim}]^2\}}{FS} \tag{32}$$

where

$$FS = \frac{5}{3} + \frac{3(\ell_e/r)}{8(\ell_e/r)_{lim}} - \frac{(\ell_e/r)^3}{8[(\ell_e/r)_{lim}]^3} \tag{33}$$

The factor of safety given by Equation (33) varies from 1.62 at $\ell_e/r = 0$ to 1.92 at the upper limit of the intermediate-column range.

Summary

1 Calculate ℓ_e/r.
2 Calculate $(\ell_e/r)_{lim}$.
3 When $\ell_e/r < (\ell_e/r)_{lim}$, use Equations (33) and (32).
   When $\ell_e/r \geq (\ell_e/r)_{lim}$, use Equation (31).

Empirical column formula for aluminum

The maximum allowable load for an aluminum column in the intermediate-column range is usually computed by a straight-line formula, such as Equation (29). The values of the constants are dependent on the particular aluminum alloy. As an example, the following equation corresponds to the rather common 2024-T4 aluminum alloy, and has been taken from a group of similar formulas in the *ALCOA Structural Handbook* (8th edition, Aluminum Company of America, Pittsburgh, Pa., 1960, p. 110), where $\sigma_0 = 44.8$ and $C_2 = 0.313$:

$$\sigma_{cr} = -\frac{P}{A} = \left[-44.8 + 0.313\left(\frac{\ell_e}{r}\right)\right] \text{ksi}$$

$$18.5 \le \frac{\ell_e}{r} \le 64 \tag{34}$$

Formula (34) does not contain any safety factor, and the *ALCOA Structural Handbook* does not suggest any specific value for it. Consequently, it is left to the practicing engineer to determine the appropriate safety factor for his or her particular problem.

Turn to Section SG10-3 of the Study Guide. ■ **STOP**

Problems

10-1.1 Let's assume that a piece of equipment is to be supported at the top of a standard steel pipe with a 5-in. nominal diameter ($d_0 = 5.563$ in., $d_i = 5.047$ in., $I = 15.16$ in^4, $A = 4.3$ in^2). The equipment and its supporting platform weigh 5000 lb. If FS = 2.0, what is the maximum L? Let $E = 28 \times 10^6$ psi.

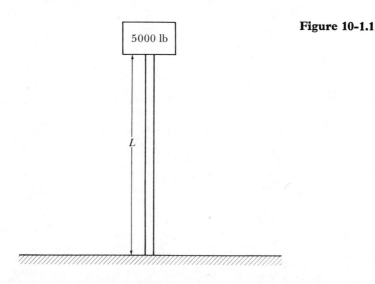

Figure 10-1.1

5000 lb

L

10-1.2 Assume that a thin bar of stainless steel is given an initial compression load of 10 lb. The bar is 8 in. long, and has an initial temperature of 75 °F. How much may the temperature of the bar rise before the bar begins to buckle? Assume that $E = 28 \times 10^6$ psi and $\alpha = 9 \times 10^{-6}/°F$.

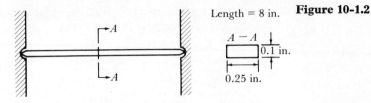

Length = 8 in.   **Figure 10-1.2**

10-1.3 The pin-connected frame shown in the figure carries a concentrated load F. Assume that buckling can occur only in the plane of the frame. Determine the maximum value of F. Assume that the material is aluminum with $E = 11.5 \times 10^6$ psi $= 79.3$ GPa. Both members have 40 mm × 40 mm cross sections.

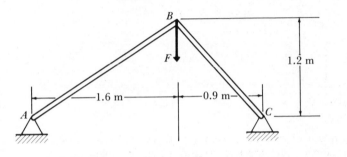

Figure 10-1.3

10-1.4 Let's assume that a 0.2%-C steel bar with a circular cross section is to be anchored firmly in the ground at one end, with the other end free. Determine the necessary radius of the bar's cross section, given that the bar is to support the weight of a man (80 kg) as he climbs to the top. The length of the bar is to be 15 m, and a safety factor of 4 is required.

10-1.5 The following frame is constructed of aluminum ($E = 79$ GPa) members. Determine the load P that will cause member AB to buckle.

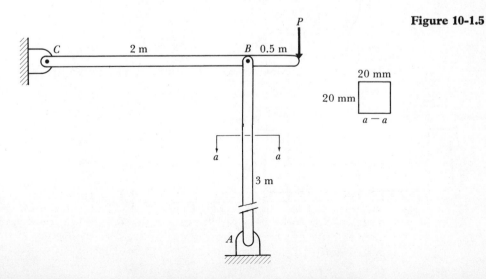

Figure 10-1.5

10-1.6 Determine the maximum allowable load for the bar shown in the figure. Assume that $E = 30 \times 10^6$ psi.

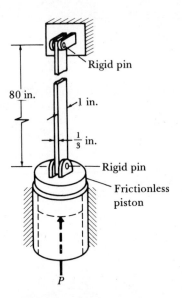

Figure 10-1.6

80 in.

Rigid pin

1 in.

$\frac{1}{3}$ in.

Rigid pin

Frictionless piston

P

10-2.1 A 2-in.-diameter steel bar is subjected to a compression load P, located as shown. The effective length for bending about the x axis is 75 in., but for bending about the y axis, the end conditions reduce the effective length to 40 in. Determine the load P that may be applied with a factor of safety of 1.8. Assume that $E = 30 \times 10^6$ psi and that $\sigma_{YP} = 40,000$ psi.

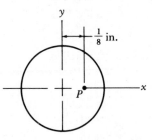

Figure 10-2.1

y

$\frac{1}{8}$ in.

P

x

10-2.2 Determine the maximum load P that can be applied to the gray-

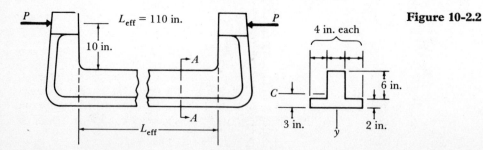

P

$L_{\text{eff}} = 110$ in.

P

Figure 10-2.2

4 in. each

10 in.

A

C

6 in.

A

L_{eff}

3 in.

y

2 in.

cast-iron frame. Since gray cast iron is a brittle material, use the ultimate compressive strength for σ_{YP}. (I_{cc} = 272 in⁴, I_{yy} = 320 in⁴)

10-2.3 Determine the maximum load P that can be applied to the following steel bar. Assume that ec/r^2 = 0.2, σ_{YP} = 40,000 psi, and E = 30 × 10^6 psi. Also compute σ_{max} given that P = $P_{max}/2$.

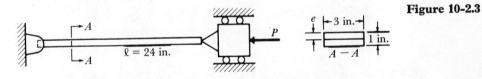

Figure 10-2.3

10-2.4 Let's examine aluminum (E = 79 GPa) bar with a circular cross section (radius = 20 mm) and a length of 0.6 m. One end of the bar is rigidly fixed, and the other end is free. Determine the maximum allowable compressive load P. Assume that ec/r^2 = 0.1 and that σ_{YP} = 276 MPa (40,000 psi).

10-2.5 Compute the maximum allowable weight W in kilograms for the following steel column. Use a safety factor of 2. Assume that σ_{YP} = 276 MPa.

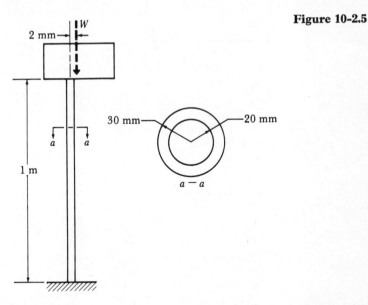

Figure 10-2.5

10-3.1 Compute the allowable weight of the equipment in Problem 10-1.1, given that L = 8 ft. Assume that σ_{YP} = 36 ksi.

10-3.2 Use the information presented in Section R10-3 to determine the maximum allowable value of F in the following pin-connected frame. Assume the material to be 2024-T4 aluminum alloy, the safety factor to be 2, and that both members have square cross sections 1.5 in. × 1.5 in.

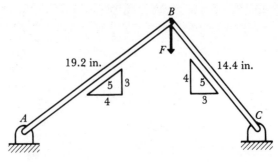

Figure 10-3.2

10-3.3 Compute the allowable load P in Problem 10-1.5, given that the cross section of member AB is changed as shown. Assume that $\sigma_{YP} = 276$ MPa.

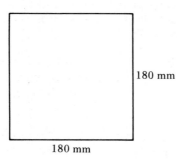

Figure 10-3.3

Appendix one

The International System of Units (SI)

The Eleventh General Conference on Weights and Measures (held in Paris in October 1960) adopted the MKSA (meter–kilogram–second–ampere) system for the whole engineering profession. The system was called "Le Système Internationale d'Unites" (The International System of Units), and given the symbol *SI* in all languages. Thirty-six countries, including the United States, participated in this conference, and recommended SI for all scientific, technical, practical, and teaching use.

SI is a simple, coherent, and rationalized selection of units from the metric system. Its great advantage results from the use of one and only one unit for each physical quantity, such as the meter for length, the kilogram for mass, the second for time, and the ampere for current.

As we describe the SI system in the following pages, you will note a difference in the way numbers are written, with spaces instead of commas. The reason is that, in some countries, commas are used in the same way we use decimal points. To avoid confusion, in the SI system we write the number 1,000,000, for example, as 1 000 000.

Definitions (official translation from the French)

The *meter* (m) is the length equal to 1 650 763.73 wavelengths in vacuum of the radiation corresponding to the transition between the levels $2p_{10}$ and $5d_5$ of the krypton-86 atom.

The *kilogram* (kg) is the unit of mass. It is equal to the mass of the international prototype of the kilogram.

The *second* (s) is the duration of 9 192 631 770 periods of the radiation corresponding to the transition between the two hyperfine levels of the ground state of the cesium-133 atom.

The *ampere* (A) is that constant current which, if maintained in two straight parallel conductors of infinite length, of negligible circular cross section, and placed 1 meter apart in vacuum, would produce between these conductors a force equal to 2×10^{-7} newton per meter of length.

The *kelvin* (K), the unit of thermodynamic temperature, is the fraction 1/273.16 of the thermodynamic temperature of the triple point of water.

The *candela* (cd) is the luminous intensity, in the perpendicular direction, of a surface of 1/600 000 square meter of a blackbody at the temperature of freezing platinum under a pressure of 101 325 newtons per square meter.

The *mole* (mol) is the amount of substance in a system containing as many elementary entities as there are atoms in 0.012 kilogram of carbon 12. When one uses the mole, the elementary entities ought to be specified. These entities can be atoms, molecules, ions, electrons, other particles, or specified groups of such particles.

The *radian* is the unit of measure of a plane angle with its vertex at the center of a circle and subtended by an arc equal in length to the radius.

The *steradian* is the unit of measure of a solid angle with its vertex at the center of a sphere and enclosing an area of the spherical surface equal to that of a square with sides equal in length to the radius.

Definitions of derived SI units having special names

Physical quantity	Unit and definition
Electric capacitance	The *farad* is the capacitance of a capacitor between the plates of which there appears a difference of potential of one volt when it is charged by a quantity of electricity equal to one coulomb.
Electric inductance	The *henry* is the inductance of a closed circuit in which an electromotive force of one volt is produced when the electric current in the circuit varies uniformly at a rate of one ampere per second.
Electric potential difference (electromotive force)	The *volt* (unit of electric potential difference and electromotive force) is the difference in electric potential between two points of a conductor carrying a constant current of one ampere, when the power dissipated between these points is equal to one watt.

Physical quantity	Unit and Definition
Electric resistance	The *ohm* is the electric resistance between two points of a conductor when a constant difference of potential of one volt, applied between these two points, produces in this conductor a current of one ampere, this conductor not being the source of any electromotive force.
Energy	The *joule* is the work done when the point of application of a force of one newton is displaced a distance of one meter in the direction of the force.
Force	The *newton* is that force which, when applied to a body having a mass of one kilogram, gives it an acceleration of one meter per second per second.
Frequency	The *hertz* is a frequency of one cycle per second.
Illumination	The *lux* is the luminous intensity given by a luminous flux of one lumen per square meter.
Luminous flux	The *lumen* is the luminous flux emitted in a solid angle of one steradian by a point source having a uniform intensity of one candela.
Magnetic flux	The *weber* is the magnetic flux which, linking a circuit of one turn, produces in it an electromotive force of one volt as it is reduced to zero at a uniform rate in one second.
Magnetic flux density	The *tesla* is the magnetic flux density given by a magnetic flux of one weber per square meter.
Power	The *watt* is the power which gives rise to the production of energy at the rate of one joule per second.
Quantity of electricity	The *coulomb* is the quantity of electricity transported in one second by a current of one ampere.

Prefixes in SI

It had originally been intended to use Greek prefixes for multiples of 10 (thus *deka, hecto,* and *kilo* for 10, 100, and 1000, respectively), and Latin prefixes for submultiples of 10 (thus *deci, centi,* and *milli* for 0.1, 0.01, and 0.001, respectively). This scheme broke down, however, when *micro,* a Greek word, was used for 10^{-6}. The following table gives all officially approved prefixes, with their accepted symbols. The prefixes *mega, giga,* and *tera* are of Greek origin, and mean large, gigantic, and monstrous, respectively. *Nano* is also of Greek origin, meaning dwarf. *Pico* is Italian for extremely small, while *femto* and *atto* are derived from the Danish numerals *femten* (fifteen) and *atten* (eighteen).

	Power	Prefix	Symbol
$1\ 000\ 000\ 000\ 000 = 10^{12}$	$= E + 12$	tera	T
$1\ 000\ 000\ 000 = 10^{9}$	$= E + 09$	giga	G
$1\ 000\ 000 = 10^{6}$	$= E + 06$	mega	M
$1\ 000 = 10^{3}$	$= E + 03$	kilo	k
$100 = 10^{2}$	$= E + 02$	hecto	h
$10 = 10^{1}$	$= E + 01$	deka	da
$0.1 = 10^{-1}$	$= E - 01$	deci	d
$0.01 = 10^{-2}$	$= E - 02$	centi	c
$0.001 = 10^{-3}$	$= E - 03$	milli	m
$0.000\ 001 = 10^{-6}$	$= E - 06$	micro	μ
$0.000\ 000\ 001 = 10^{-9}$	$= E - 09$	nano	n
$0.000\ 000\ 000\ 001 = 10^{-12}$	$= E - 12$	pico	p
$0.000\ 000\ 000\ 000\ 001 = 10^{-15}$	$= E - 15$	femto	f
$0.000\ 000\ 000\ 000\ 000\ 001 = 10^{-18}$	$= E - 18$	atto	a

Other units associated with SI

The *liter* was originally (1795) intended to be identical to the cubic decimeter (1 000 cm^3). Subsequently (1901) it was decided to define the liter as the volume occupied by the mass of one kilogram of pure water at its maximum density under normal atmospheric pressure. Exact measurements later showed, however, that one kilogram of water as defined above occupies a volume 1.000 028 dm^3. This is why the present official definition of the liter is back to 1 liter = 1 cubic decimeter; it is not officially defined in terms of the mass of water. The liter is commonly used in metric countries as a measure of volume, but it is not an official SI unit.

The *bar*, a unit of pressure, is approximately equal to one atmosphere (14.5 psi) and exactly equal to 100 kilonewtons per square meter. It is used extensively by meteorologists (it is also a root word for pressure-related terms such as "isobar" and "barometer"), but again, it is not an official SI unit.

The *pascal* (Pa) is an acceptable name for the unit of pressure or stress; it is equivalent to newton per square meter (1 Pa = 1 N/m^2).

The *siemens* (S) is also acceptable as the name for the unit of electrical conductance. It is equivalent to the reciprocal ohm or the ampere per volt (1 S = 1 A/V).

Mass, weight, force

One of the outstanding advantages of SI over any other system of measurement is the fact that mass is defined as a base unit (unit =

kilogram), while force or weight is determined from $F = mg$. Thus a clear distinction is made between mass and force, which eliminates much of the confusion resulting from the use of other systems. The unit of force is appropriately called the *newton* (Newton's second law: $F = ma$). One newton (N) is defined as "the force required to impart an acceleration of one meter per second per second to a mass of one kilogram." Since the standard acceleration of gravity on earth is

$$g_0 = 9.806\ 65\ \text{m/s}^2$$

the relationship between units of mass and weight (or force) for most engineering and scientific purposes can be given as

Force = weight = 9.806 65 times mass

where the mass is in kilograms and F (or W) is the force in newtons. For engineering computations, $F = 9.8$ times mass will be sufficiently accurate.

The correct use of the terms mass and weight is the following: The term *mass* is used to specify the quantity of matter contained in material objects. Mass is independent of location in the universe.

The term *weight* is used as a measure of the gravitational force acting on a material object at a specified location; it generally varies as the object changes location. Even on the earth's surface, the gravitational force varies by about 0.5%. This variation must be taken into consideration where great accuracy is required.

Rules for SI style

1 It is recommended that approved prefixes be used to indicate orders of magnitude. For example,

15 200 m or 15.2×10^3 m becomes 15.2 km

and

125 000 000 N or 125×10^6 N becomes 125 MN

2 Symbols for SI units are not capitalized unless the unit is derived from a proper name. Thus we use m for meter, but N for newton. Only the numerical prefixes T (tera), G (giga), and M (mega) are capitalized.
3 SI symbols are always written in singular form. Thus 215 newtons = 215 N and 85 kilometers = 85 km. Periods should not be used after SI unit symbols (except at the end of a sentence).
4 To avoid confusion caused by the use of commas in some countries to express decimal points, numbers having four or more digits should be placed in groups of three separated by a space. For example, use

1 253 725   instead of   1,253,725

and

3 283.025 72   instead of   3,283.02572

Rules for conversion and rounding

In all conversions, the number of significant digits retained should be such that the implied or stated accuracy is neither sacrificed nor exaggerated. The most accurate equivalents are obtained by multiplying the specified quantity by the most accurate conversion factor available, and then rounding to the appropriate number of significant digits.

The number of significant digits to be retained after multiplication, division, addition, or subtraction is determined as follows.

A product or quotient shall contain no more significant digits than are contained in the number with the fewest significant digits used in the multiplication or division. For example,

$113.2 \times 1.43 = 161.876$ must be rounded to 162

$113.2 \div 1.43 = 79.161$ must be rounded to 79.2

The answer of an addition or subtraction shall contain no significant digits farther to the right than occurs in the least accurate figure. For example, assume that in the following three numbers to be added, the zeros do not indicate a specific value, but only the magnitude of the number.

$$
\begin{array}{r}
251\ 000 \\
132\ 780 \\
4\ 762 \\
\hline
388\ 542
\end{array}
$$

This total implies an unrealistic precision. The numbers should first be rounded to one significant digit farther to the right than that of the least accurate number, and then added, as follows.

$$
\begin{array}{r}
251\ 000 \\
132\ 800 \\
4\ 800 \\
\hline
388\ 600
\end{array}
$$

The answer is then rounded to 389 000.

Reference 1 presents a more comprehensive treatment of the rules for conversion and rounding.

SI units in mechanics of materials

General principles of calculation One of the main reasons for the establishment of SI is that its correct use greatly simplifies calculation. The advantages derived from the decimal nature of the metric system are obvious. But the advantages of using SI are even greater, because it is a "coherent" system. "Coherent" means that all units of the system are either base units or units derived from base units by simply multiplying or dividing, without using any factor other than unity. This means that if one uses the correct base units and derived units in calculations, the result will be expressed in the correct unit. This constitutes a considerable advantage over the old English system of units, and even over the old metric systems.

In order to derive maximum benefit from the advantages of SI, one must use an orderly, systematic approach to all problem solving. The following procedure is recommended.

1 Express all available data in terms of SI base or derived units, and express prefixes in terms of powers of 10.
2 Make sure that the formula to be used is expressed in SI base or derived units, and not in multiples or submultiples of them.
3 Substitute data into formulas. The result will be in terms of a SI base or derived unit.
4 Express the result in a convenient and acceptable multiple or submultiple of the unit.

The following are the SI base and derived units that are most commonly encountered in problems of mechanics of materials.

SI base units

Quantity	Name	Symbol
Length	meter	m
Mass	kilogram	kg
Time	second	s

SI derived units

Quantity	Name	Symbol	Definition
Energy	joule	J	$N \cdot m$
Force	newton	N	$kg \cdot m/s^2$
Frequency	hertz	Hz	s^{-1}
Power	watt	W	J/s
Pressure	pascal	Pa	N/m^2
Stress	pascal	Pa	N/m^2
Weight	newton	N	$kg \cdot m/s^2$
Work	joule	J	$N \cdot m$

The following are common mechanics-of-materials formulas in correct SI form.

Average normal stress in centrically loaded bar

$$\sigma = \frac{P}{A} = \frac{mg}{A}$$

where

σ = stress in newtons per square meter (N/m^2) or in pascals (Pa)
P = force in newtons (N)
m = mass in kilograms (kg)
g = acceleration of gravity in meters per second per second

$$(g = 9.806\ 65\ \text{m/s}^2 \quad \text{or} \quad g \cong 9.8\ \text{m/s}^2)$$

A = cross-sectional area of bar in square meters (m^2)

Deformation of centrically loaded bar

$$\delta = \frac{PL}{AE} = \frac{mgL}{AE}$$

where

δ = deformation of bar in meters (m)
L = length of bar in meters (m)
E = modulus of elasticity (N/m^2 or Pa)

Torsion of cylindrical bar

$$\tau = \frac{T\rho}{J}$$

where

τ = shearing stress (N/m^2 or Pa)
T = torque in newton-meters (N·m)
ρ = radial distance from center of bar (m)
J = polar moment of inertia (m^4)

Angular deformation of torsionally loaded bar

$$\theta = \frac{TL}{JG}$$

where

θ = angular deformation in radians (rad)
L = length of bar (m)
G = modulus of rigidity (N/m^2 or Pa)
J = polar moment of inertia (m^4)

Power transmission

$$P = T\omega$$

where

P = power transmitted in watts (W)
T = torque in shaft (N·m)
ω = angular velocity in radians/second (rad/s)

Flexure of beams

$$\sigma = \frac{My}{I}$$

where

$M =$ bending moment in newton-meters (N·m)
$y =$ distance from centroidal axis (m)
$I =$ centroidal moment of inertia (m^4)

Other relationships in flexure of beams

$$w = \frac{dV}{dx}$$

$$V = \frac{dM}{dx}$$

$$M = EI \frac{d^2y}{dx^2}$$

where

$V =$ shearing force in newtons (N)
$w =$ distributed load on beam in newtons per meter (N/m)
$y =$ deflection of beam in meters (m)

Strain energy in centrically loaded bar

$$U = \frac{P^2L}{2AE} = \frac{(mg)^2L}{2AE} = \frac{AE\delta^2}{2L}$$

where

$U =$ strain energy in joules (J) (1 J = 1 N·m); all other terms as defined previously

Strain energy in torsionally loaded cylindrical bar

$$U = \frac{T^2L}{2GJ} = \frac{GJ\theta^2}{2L}$$

All terms have been defined above.

Strain energy in beams

$$U = \int_0^L \frac{M^2 \, dx}{2EI} = \int_0^L \frac{EI}{2} \left(\frac{d^2y}{dx^2}\right)^2 dx$$

All terms have been defined above.

EXAMPLE   The following example will illustrate the recommended procedure for solving problems.

A 5000-kg mass is suspended from a straight, homogeneous steel bar as shown in Figure A1-1. The dimensions of the bar are 20 mm by 30 mm by 185 cm. The modulus of elasticity is 200 GPa, and the acceleration of gravity may be taken as 9.8 m/s^2. Determine the average normal stress in the bar and its elongation.

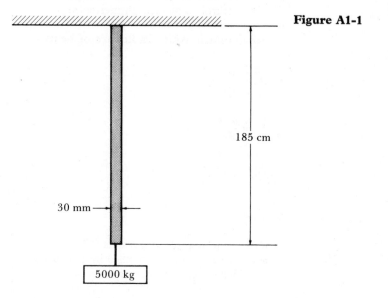

Figure A1-1

185 cm

30 mm

5000 kg

SOLUTION

Data		Expressed in SI units
mass	= 5000 kg	5.0 × 10^3 kg
width	= 20 mm	2.0 × 10^{-2} m
depth	= 30 mm	3.0 × 10^{-2} m
length	= 185 cm	1.85 m
E	= 200 GPa	200 × 10^9 N/m^2

Required: stress (σ) and elongation (δ)

Equations

$$\sigma = \frac{P}{A} = \frac{mg}{A} \quad \text{and} \quad \delta = \frac{PL}{AE} = \frac{mgL}{AE}$$

Calculations

$$\sigma = \frac{5.0 \times 10^3 \times 9.8}{(2.0 \times 10^{-2})(3.0 \times 10^{-2})} = 81.67 \times 10^6 \text{ N/m}^2$$

$$= 82 \text{ MN/m}^2 \quad \text{or} \quad 82 \text{ MPa}$$

$$\delta = \frac{5.0 \times 10^3 \times 9.8 \times 1.85}{(2.0 \times 10^{-2})(3.0 \times 10^{-2})(200 \times 10^9)} = 7.554 \times 10^{-4} \text{ m}$$

$$= 0.76 \text{ mm}$$

Some common properties of materials (Approximate values)

Modulus of elasticity (10^9 N/m^2 = 1 GN/m^2 = 1 kN/mm^2 = 1 GPa)	
Steel	200 to 210 GPa
Aluminum alloy	70 to 80 GPa
Copper alloy	90 to 115 GPa
Brass	90 GPa

Density (kg/m^3)	
Water	1000 kg/m^3
Steel	7850 kg/m^3
Aluminum alloy	2700 kg/m^3
Copper alloy	8600 kg/m^3
Air (at sea level)	1.23 kg/m^3

Weight (force) per unit volume (on earth)	
Water	9.8 kN/m^3
Steel	77 kN/m^3
Concrete	23 kN/m^3

Approximate values of some common quantities

Absolute zero	0 K = −273°C
Normal temperature of human body	37°C
Diameter of earth	12 740 km
Speed of light	300 Mm/s
Speed of sound (at sea level)	341 m/s
Acceleration due to gravity	9.81 m/s

Temperature conversion

From	To	Formula
degree Celsius (C)	Kelvin (K)	$t_K = t_C + 273.15$
degree Fahrenheit (F)	Kelvin (K)	$t_K = (t_F + 459.67)/1.8$
degree Fahrenheit (F)	degree Celsius (C)	$t_C = (t_F - 32)/1.8$

Given British system	Multiply by	To convert to MKS system	Conversion factors
inch	2.54×10^{-2}	meter (m)	
in^2	$6.451\,6 \times 10^{-4}$	m^2	
in^3	$1.638\,706 \times 10^{-5}$	m^3	
foot	3.048×10^{-1}	meter (m)	
cfm	$4.719\,474 \times 10^{-4}$	m^3/s	
gallon	$3.785\,412 \times 10^{-3}$	m^3	
gpm	$6.309\,02 \times 10^{-5}$	m^3/s	
lb$_m$	$4.535\,924 \times 10^{-1}$	kilogram (kg)	
lb$_m$/in^3	$2.767\,991 \times 10^{4}$	kg/m^3	
lb$_m$/ft^3	$1.601\,8 \times 10^{1}$	kg/m^3	
ft^3/lb$_m$	$6.242\,797 \times 10^{-2}$	m^3/kg	
lb$_f$	$4.448\,222$	newton (N)	
lb$_f$/in.	$1.751\,286 \times 10^{2}$	N/m	
lb$_f$/ft	$1.459\,39 \times 10^{1}$	N/m	
lb$_f$/in^2	$6.894\,757 \times 10^{3}$	N/m^2	
lb$_f$-foot	$1.355\,818$	newton-meter (N·m)	
lb$_f$-in.	$1.112\,985 \times 10^{-1}$	N·m	
horsepower	$7.456\,999 \times 10^{2}$	watt (W)	
inch of mercury (60°F)	$3.376\,85 \times 10^{3}$	N/m^2	
inch of water (60°F)	$2.488\,4 \times 10^{2}$	N/m^2	
mile/hour	$1.609\,344$	km/hr	
BTU	$1.055\,06 \times 10^{3}$	joule (J)	
BTU/lb	$2.326\,009 \times 10^{3}$	J/kg	
degree Rankine	4.619×10^{3}	J/kmol	
degree Kelvin	$8.314\,3 \times 10^{3}$	J/kmol	

Physical constants

Velocity of light in vacuum (c) = $2.997\,925 \pm 0.000\,003 \times 10^{8}$ m/s
Gravitational constant $C = 6.670 \pm 0.015 \times 10^{-11}$ N·m/kg^2
Standard acceleration of gravity (g_0) = $9.806\,65$ m/s^2
Normal atmospheric pressure (Pa) = $10.132\,5 \times 10^{4}$ N/m^2

References

Metric Practice Guide, E380-72, American Society for Testing and Materials
SI-Metric, IBM Reference Manual
"American National Standard Practice for Inch-Millimeter Conversion for Industrial Use," ANSI B48.1-1933 (R1947), ISO R370-1964

Appendix two

Mechanical properties of materials and properties of standard cross-sectional areas

Table 1 Average properties of selected engineering materials

Exact values may vary widely with changes in composition, heat treatment, and mechanical working. More precise information can be obtained from manufacturers.

Materials	Specific weight lb/in³	Elastic strength* Tension ksi	Elastic strength* Comp. ksi	Elastic strength* Shear ksi	Ultimate strength Tension ksi	Ultimate strength Comp. ksi	Ultimate strength Shear ksi	Modulus of elasticity 1000 ksi	Modulus of rigidity 1000 ksi	Coefficient of thermal expansion 10⁻⁸ in. per in. per °F
Ferrous metals										
Wrought iron	0.278	30	**		48	**	25	28		6.7
Structural steel	0.284	36	**		66	**		29	11.0	6.6
Steel, 0.2% C, hardened	0.284	62	**		90	**		30	11.6	6.6
Steel, 0.4% C, hot-rolled	0.284	53	**		84	**		30	11.6	
Steel, 0.8% C, hot-rolled	0.284	76	**		122	**		30	11.6	
Cast iron, gray	0.260				25	100		15		6.7
Cast iron, malleable	0.266	32	**		50	**		25		6.6
Cast iron, nodular	0.266	70			100			25		6.6
Stainless steel (18-8), annealed	0.286	36	**		85	**		28	12.5	9.6
Stainless steel (18-8), cold-rolled	0.286	165	**		190	**	95	28	12.5	
Steel, SAE 4340, heat-treated	0.283	132	145		150	**		29	11.0	9.6
Nonferrous metal alloys										
Aluminum, cast, 195-T6	0.100	24	25		36		30	10.3	3.8	
Aluminum, wrought, 2014-T4	0.101	41	41	24	62	**	38	10.6	4.0	12.5
Aluminum, wrought, 2024-T4	0.100	48	48	28	68	**	41	10.6	4.0	12.5
Aluminum, wrought, 6061-T6	0.098	40	40	26	45	**	30	10.0	3.8	12.5
Magnesium, extruded, AZ80X	0.066	35	26		49	**	21	6.5	2.4	14.4
Magnesium, sand-cast, AZ63-HT	0.066	14	14		40	**	19	6.5	2.4	14.4

Monel, wrought, hot-rolled	0.319	50	**	90	**		26	9.5	7.8
Red brass, cold-rolled	0.316	60		75			15	5.6	9.8
Red brass, annealed	0.316	15	**	40	**		15	5.6	9.8
Bronze, cold-rolled	0.320	75		100			15	6.5	9.4
Bronze, annealed	0.320	20	**	50	**			6.5	9.4
Titanium alloy, annealed	0.167	135	**	155	**		14	5.3	
Invar, annealed	0.292	42	**	70	**		21	8.1	0.6
Nonmetallic materials									
Douglas fir, green	0.022	4.8	3.4		3.9	0.9	1.6		
Douglas fir, air-dried	0.020	8.1	6.4		7.4	1.1	1.9		
Red oak, green	0.037	4.4	2.6		3.5	1.2	1.4		1.9
Red oak, air-dried	0.025	8.4	4.6		6.9	1.8	1.8		
Concrete, medium strength	0.087		1.2		3.0		3.0		6.0
Concrete, fairly high strength	0.087		2.0		5.0		4.5		6.0

* Elastic strength may be represented by proportional limit, yield point, or yield strength at a specified offset (usually 0.2% for ductile metals).

** For ductile metals (those that have an appreciable ultimate elongation), it is customary to assume that the properties in compression have the same values as those in tension.

All timber properties are taken parallel to the grain.

Courtesy of John Wiley & Sons, *Mechanics of Materials*, Higdon, Ohlsen, Stiles, and Weese, 1967

Table 2 American standard: steel I-beams (Properties for designing)

Designation	Area in²	Depth in.	Flange Width in.	Flange Thickness in.	Web thickness in.	Axis xx I in⁴	Axis xx S in³	Axis xx r in.	Axis yy I in⁴	Axis yy S in³	Axis yy r in.
S 24 × 120.0	35.3	24.00	8.048	1.102	0.798	3030	252	9.26	84.2	20.9	1.54
× 105.9	31.1	24.00	7.875	1.102	0.625	2830	236	9.53	78.2	19.8	1.58
S 24 × 100.0	29.4	24.00	7.247	0.871	0.747	2390	199	9.01	47.8	13.2	1.27
× 90.0	26.5	24.00	7.124	0.871	0.624	2250	187	9.22	44.9	12.6	1.30
× 79.9	23.5	24.00	7.001	0.871	0.500	2110	175	9.47	42.3	12.1	1.34
S 20 × 95.0	27.9	20.00	7.200	0.916	0.800	1610	161	7.60	49.7	13.8	1.33
× 85.0	25.0	20.00	7.053	0.916	0.653	1520	152	7.79	46.2	13.1	1.36
S 20 × 75.0	22.1	20.00	6.391	0.789	0.641	1280	128	7.60	29.6	9.28	1.16
× 65.4	19.2	20.00	6.250	0.789	0.500	1180	118	7.84	27.4	8.77	1.19
S 18 × 70.0	20.6	18.00	6.251	0.691	0.711	926	103	6.71	24.1	7.72	1.08
× 54.7	16.1	18.00	6.001	0.691	0.460	804	89.4	7.07	20.8	6.94	1.14
S 15 × 50.0	14.7	15.00	5.640	0.622	0.550	486	64.8	5.75	15.7	5.57	1.03
× 42.9	12.6	15.00	5.501	0.622	0.411	447	59.6	5.95	14.4	5.23	1.07
S 12 × 50.0	14.7	12.00	5.477	0.659	0.687	305	50.8	4.55	15.7	5.74	1.03
× 40.8	12.0	12.00	5.252	0.659	0.462	272	45.4	4.77	13.6	5.16	1.06
S 12 × 35.0	10.3	12.00	5.078	0.544	0.428	229	38.2	4.72	9.87	3.89	0.980
× 31.8	9.35	12.00	5.000	0.544	0.350	218	36.4	4.83	9.36	3.74	1.00
S 10 × 35.0	10.3	10.00	4.944	0.491	0.594	147	29.4	3.78	8.36	3.38	0.901
× 25.4	7.46	10.00	4.661	0.491	0.311	124	24.7	4.07	6.79	2.91	0.954
S 8 × 23.0	6.77	8.00	4.171	0.425	0.441	64.9	16.2	3.10	4.31	2.07	0.798
× 18.4	5.41	8.00	4.001	0.425	0.271	57.6	14.4	3.26	3.73	1.86	0.831
S 7 × 20.0	5.88	7.00	3.860	0.392	0.450	42.4	12.1	2.69	3.17	1.64	0.734
× 15.3	4.50	7.00	3.662	0.392	0.252	36.7	10.5	2.86	2.64	1.44	0.766
S 6 × 17.25	5.07	6.00	3.565	0.359	0.465	26.3	8.77	2.28	2.31	1.30	0.675
× 12.5	3.67	6.00	3.332	0.359	0.232	22.1	7.37	2.45	1.82	1.09	0.705
S 5 × 14.75	4.34	5.00	3.284	0.326	0.494	15.2	6.09	1.87	1.67	1.01	0.620
× 10.0	2.94	5.00	3.004	0.326	0.214	12.3	4.92	2.05	1.22	0.809	0.643
S 4 × 9.5	2.79	4.00	2.796	0.293	0.326	6.79	3.39	1.56	0.903	0.646	0.569
× 7.7	2.26	4.00	2.663	0.293	0.193	6.08	3.04	1.64	0.764	0.574	0.581
S 3 × 7.5	2.21	3.00	2.509	0.260	0.349	2.93	1.95	1.15	0.586	0.468	0.516
× 5.7	1.67	3.00	2.330	0.260	0.170	2.52	1.68	1.23	0.455	0.390	0.522

Courtesy of American Institute of Steel Construction

Table 3 Steel, wide-flange beams*
Properties for designing; abridged list

Designation	Area in²	Depth in.	Width in.	Flange Thickness in.	Web thickness in.	Axis xx I in⁴	S in³	r in.	Axis yy I in⁴	S in³	r in.
W 36 × 230	67.7	35.88	16.471	1.260	0.761	15000.0	837.0	14.9	940.0	114.0	3.73
× 150	44.2	35.84	11.972	0.940	0.625	9030.0	504.0	14.3	270.0	45.0	2.47
W 33 × 200	58.9	33.00	15.750	1.150	0.715	11100.0	671.0	13.7	750.0	95.2	3.57
× 130	38.3	33.10	11.510	0.855	0.580	6710.0	406.0	13.2	218.0	37.9	2.38
W 30 × 172	50.7	29.88	14.985	1.065	0.655	7910.0	530.0	12.5	598.0	79.8	3.43
× 108	31.8	29.82	10.484	0.760	0.548	4470.0	300.0	11.9	146.0	27.9	2.15
W 27 × 145	42.7	26.88	13.965	0.975	0.600	5430.0	404.0	11.3	443.0	63.5	3.22
× 94	27.7	26.91	9.990	0.747	0.490	3270.0	243.0	10.9	124.0	24.9	2.12
W 24 × 130	38.3	24.25	14.000	0.900	0.565	4020.0	332.0	10.2	412.0	58.9	3.28
× 100	29.5	24.00	12.000	0.775	0.468	3000.0	250.0	10.1	223.0	37.2	2.75
× 76	22.4	23.91	8.985	0.682	0.440	2100.0	176.0	9.69	82.6	18.4	1.92
W 21 × 112	33.0	21.00	13.000	0.865	0.527	2620.0	250.0	8.92	317.0	48.8	3.10
× 82	24.2	20.86	8.962	0.795	0.499	1760.0	169.0	8.53	95.6	21.3	1.99
× 62	18.3	20.99	8.240	0.615	0.400	1330.0	127.0	8.54	57.5	13.9	1.77
W 18 × 96	28.2	18.16	11.750	0.831	0.512	1680.0	185.0	7.70	225.0	38.3	2.82
× 64	18.9	17.87	8.715	0.686	0.403	1050.0	118.0	7.46	75.8	17.4	2.00
× 50	14.7	18.00	7.500	0.570	0.358	802.0	89.1	7.38	40.2	10.7	1.65
W 16 × 88	25.9	16.16	11.502	0.795	0.504	1220.0	151.0	6.87	202.0	35.1	2.79
× 58	17.1	15.86	8.464	0.645	0.407	748.0	94.4	6.62	65.3	15.4	1.96
× 50	14.7	16.25	7.073	0.628	0.380	657.0	80.8	6.68	37.1	10.5	1.59
× 36	10.6	15.85	6.992	0.428	0.299	447.0	56.5	6.50	24.4	6.99	1.52
W 14 × 142	41.8	14.75	15.500	1.063	0.680	1670.0	227.0	6.32	660.0	85.2	3.97
× 320	94.1	16.81	16.710	2.093	1.890	4140.0	493.0	6.63	1640.0	196.0	4.17
× 87	25.6	14.00	14.500	0.688	0.420	967.0	138.0	6.15	350.0	48.2	3.70

* Frequently in the literature, steel WF beams are designated by giving their nominal depth in inches first; then the letters WF to designate wide-flange beam; then the weight in pounds per linear foot. For example, 36 WF 230.

Courtesy of American Institute of Steel Construction

(Table continues on pages 230–231.)

Table 3 Continued

Designation	Area in²	Depth in.	Flange Width in.	Flange Thickness in.	Web thickness in.	Axis xx I in⁴	S in³	r in.	Axis yy I in⁴	S in³	r in.
W 14 × 84	24.7	14.18	12.023	0.778	0.451	928.0	131.0	6.13	225.0	37.5	3.02
× 78	22.9	14.06	12.000	0.718	0.428	851.0	121.0	6.09	207.0	34.5	3.00
× 74	21.8	14.19	10.072	0.783	0.450	797.0	112.0	6.05	133.0	26.5	2.48
× 68	20.0	14.06	10.040	0.718	0.418	724.0	103.0	6.02	121.0	24.1	2.46
× 61	17.9	13.91	10.000	0.643	0.378	641.0	92.2	5.98	107.0	21.5	2.45
× 53	15.6	13.94	8.062	0.658	0.370	542.0	77.8	5.90	57.5	14.3	1.92
× 43	12.6	13.68	8.000	0.528	0.308	429.0	62.7	5.82	45.1	11.3	1.89
× 38	11.2	14.12	6.776	0.513	0.313	386.0	54.7	5.88	26.6	7.86	1.54
× 34	10.0	14.00	6.750	0.453	0.287	340.0	48.6	5.83	23.3	6.89	1.52
× 30	8.83	13.86	6.733	0.383	0.270	290.0	41.9	5.74	19.5	5.80	1.49
W 12 × 85	25.0	12.50	12.105	0.796	0.495	723.0	116.0	5.38	235.0	38.9	3.07
× 65	19.1	12.12	12.000	0.606	0.390	533.0	88.0	5.28	175.0	29.1	3.02
× 53	15.6	12.06	10.000	0.576	0.345	426.0	70.7	5.23	96.1	19.2	2.48
× 40	11.8	11.94	8.000	0.516	0.294	310.0	51.9	5.13	44.1	11.0	1.94
× 36	10.6	12.24	6.565	0.540	0.305	281.0	46.0	5.15	25.5	7.77	1.55
× 31	9.13	12.09	6.525	0.465	0.265	239.0	39.5	5.12	21.6	6.61	1.54
× 27	7.95	11.95	6.497	0.400	0.237	204.0	34.2	5.07	18.3	5.63	1.52
W 10 × 112	32.9	11.38	10.415	1.248	0.755	719.0	126.0	4.67	235.0	45.2	2.67
× 100	29.4	11.12	10.345	1.118	0.685	625.0	112.0	4.61	207.0	39.9	2.65
× 89	26.2	10.88	10.275	0.998	0.615	542.0	99.7	4.55	181.0	35.2	2.63

× 77	22.7	10.62	10.195	0.868	0.535	457.0	86.1	4.49	153.0	30.1	2.60
× 49	14.4	10.00	10.000	0.558	0.340	273.0	54.6	4.35	93.0	18.6	2.54
× 45	13.2	10.12	8.022	0.618	0.350	249.0	49.1	4.33	53.2	13.3	2.00
× 39	11.5	9.94	7.990	0.528	0.318	210.0	42.2	4.27	44.9	11.2	1.98
× 33	9.71	9.75	7.964	0.433	0.292	171.0	35.0	4.20	36.5	9.16	1.94
× 29	8.54	10.22	5.799	0.500	0.289	158.0	30.8	4.30	16.3	5.61	1.38
× 21	6.20	9.90	5.750	0.340	0.240	107.0	21.5	4.15	10.8	3.75	1.32
W 8 × 67	19.7	9.00	8.287	0.933	0.575	272.0	60.4	3.71	88.6	21.4	2.12
× 58	17.1	8.75	8.222	0.808	0.510	227.0	52.0	3.65	74.9	18.2	2.10
× 48	14.1	8.50	8.117	0.683	0.405	184.0	43.2	3.61	60.9	15.0	2.08
× 40	11.8	8.25	8.077	0.558	0.365	146.0	35.5	3.53	49.0	12.1	2.04
× 35	10.3	8.12	8.027	0.493	0.315	126.0	31.1	3.50	42.5	10.6	2.03
× 31	9.12	8.00	8.000	0.433	0.288	110.0	27.4	3.47	37.0	9.24	2.01
× 28	8.23	8.06	6.540	0.463	0.285	97.8	24.3	3.45	21.6	6.61	1.62
× 24	7.06	7.93	6.500	0.398	0.245	82.5	20.8	3.42	18.2	5.61	1.61
× 20	5.89	8.14	5.268	0.378	0.248	69.4	17.0	3.43	9.22	3.50	1.25
× 17	5.01	8.00	5.250	0.308	0.230	56.6	14.1	3.36	7.44	2.83	1.22

Courtesy of American Institute of Steel Construction

Table 4 American standard: steel channels
Properties for designing

Designation	Area in²	Depth in.	Flange Width in.	Flange Average thickness in.	Web thickness in.	Axis xx I in⁴	Axis xx S in³	Axis xx r in.	Axis yy I in⁴	Axis yy S in³	Axis yy r in.	x in.
15 × 50.0	14.7	15.00	3.716	0.650	0.716	404	53.8	5.24	11.0	3.78	0.867	.80
× 40.0	11.8	15.00	3.520	0.650	0.520	349	46.5	5.44	9.23	3.36	0.886	.78
× 33.9	9.96	15.00	3.400	0.650	0.400	315	42.0	5.62	8.13	3.11	0.904	.79
12 × 30.0	8.82	12.00	3.170	0.501	0.510	162	27.0	4.29	5.14	2.06	0.763	.67
× 25.0	7.35	12.00	3.047	0.501	0.387	144	24.1	4.43	4.47	1.88	0.780	.67
× 20.7	6.09	12.00	2.942	0.501	0.280	129	21.5	4.61	3.88	1.73	0.799	.70
10 × 30.0	8.82	10.00	3.033	0.436	0.673	103	20.7	3.42	3.94	1.65	0.669	.65
× 25.0	7.35	10.00	2.886	0.436	0.526	91.2	18.2	3.52	3.36	1.48	0.676	.62
× 20.0	5.88	10.00	2.739	0.436	0.379	78.9	15.8	3.66	2.81	1.32	0.691	.61
× 15.3	4.49	10.00	2.600	0.436	0.240	67.4	13.5	3.87	2.28	1.16	0.713	.64
9 × 20.0	5.88	9.00	2.648	0.413	0.448	60.9	13.5	3.22	2.42	1.17	0.642	.58
× 15.0	4.41	9.00	2.485	0.413	0.285	51.0	11.3	3.40	1.93	1.01	0.661	.59
× 13.4	3.94	9.00	2.433	0.413	0.233	47.9	10.6	3.49	1.76	0.962	0.668	.60
8 × 18.75	5.51	8.00	2.527	0.390	0.487	44.0	11.0	2.82	1.98	1.01	0.599	.57
× 13.75	4.04	8.00	2.343	0.390	0.303	36.1	9.03	2.99	1.53	0.853	0.615	.55
× 11.5	3.38	8.00	2.260	0.390	0.220	32.6	8.14	3.10	1.32	0.781	0.625	.57
7 × 14.75	4.33	7.00	2.299	0.366	0.419	27.2	7.78	2.51	1.38	0.779	0.564	.53
× 12.25	3.60	7.00	2.194	0.366	0.314	24.2	6.93	2.60	1.17	0.702	0.571	.53
× 9.8	2.87	7.00	2.090	0.366	0.210	21.3	6.08	2.72	0.968	0.625	0.581	.54
6 × 13.0	3.83	6.00	2.157	0.343	0.437	17.4	5.80	2.13	1.05	0.642	0.525	.51
× 10.5	3.09	6.00	2.034	0.343	0.314	15.2	5.06	2.22	0.865	0.564	0.529	.50
× 8.2	2.40	6.00	1.920	0.343	0.200	13.1	4.38	2.34	0.692	0.492	0.537	.51
5 × 9.0	2.64	5.00	1.885	0.320	0.325	8.90	3.56	1.83	0.632	0.449	0.489	.48
× 6.7	1.97	5.00	1.750	0.320	0.190	7.49	3.00	1.95	0.478	0.378	0.493	.48
4 × 7.25	2.13	4.00	1.721	0.296	0.321	4.59	2.29	1.47	0.432	0.343	0.450	.46
× 5.4	1.59	4.00	1.584	0.296	0.184	3.85	1.93	1.56	0.319	0.283	0.449	.46
3 × 6.0	1.76	3.00	1.596	0.273	0.356	2.07	1.38	1.08	0.305	0.268	0.416	.46
× 5.0	1.47	3.00	1.498	0.273	0.258	1.85	1.24	1.12	0.247	0.233	0.410	.44
× 4.1	1.21	3.00	1.410	0.273	0.170	1.66	1.10	1.17	0.197	0.202	0.404	.44

Table 5 Steel angles: equal legs
Properties for designing

| Size and thickness | Weight per foot | Area | Axis xx and Axis yy | | | | Axis zz |
| | | | I | I/C | r | x or y | r |
in.	lb	in^2	in^4	in^3	in.	in.	in.
L8 × 8 × 1⅛	56.9	16.73	98.0	17.5	2.42	2.41	1.56
1	51.0	15.00	89.0	15.8	2.44	2.37	1.56
⅞	45.0	13.23	79.6	14.0	2.45	2.32	1.57
¾	38.9	11.44	69.7	12.2	2.47	2.28	1.57
⅝	32.7	9.61	59.4	10.3	2.49	2.23	1.58
9/16	29.6	8.68	54.1	9.3	2.50	2.21	1.58
½	26.4	7.75	48.6	8.4	2.50	2.19	1.59
L6 × 6 × 1	37.4	11.00	35.5	8.6	1.80	1.86	1.17
⅞	33.1	9.73	31.9	7.6	1.81	1.82	1.17
¾	28.7	8.44	28.2	6.7	1.83	1.78	1.17
⅝	24.2	7.11	24.2	5.7	1.84	1.73	1.18
9/16	21.9	6.43	22.1	5.1	1.85	1.71	1.18
½	19.6	5.75	19.9	4.6	1.86	1.68	1.18
7/16	17.2	5.06	17.7	4.1	1.87	1.66	1.19
⅜	14.9	4.36	15.4	3.5	1.88	1.64	1.19
5/16	12.5	3.66	13.0	3.0	1.89	1.61	1.19
L5 × 5 × ⅞	27.2	7.98	17.8	5.2	1.49	1.57	0.97
¾	23.6	6.94	15.7	4.5	1.51	1.52	0.97
⅝	20.0	5.86	13.6	3.9	1.52	1.48	0.98
½	16.2	4.75	11.3	3.2	1.54	1.43	0.98
7/16	14.3	4.18	10.0	2.8	1.55	1.41	0.98
⅜	12.3	3.61	8.7	2.4	1.56	1.39	0.99
5/16	10.3	3.03	7.4	2.0	1.57	1.37	0.99
L4 × 4 × ¾	18.5	5.44	7.7	2.8	1.19	1.27	0.78
⅝	15.7	4.61	6.7	2.4	1.20	1.23	0.78
½	12.8	3.75	5.6	2.0	1.22	1.18	0.78
7/16	11.3	3.31	5.0	1.8	1.23	1.16	0.78
⅜	9.8	2.86	4.4	1.5	1.23	1.14	0.79
5/16	8.2	2.40	3.7	1.3	1.24	1.12	0.79
¼	6.6	1.94	3.0	1.1	1.25	1.09	0.80
L3½ × 3½ × ½	11.1	3.25	3.6	1.5	1.06	1.06	0.68
7/16	9.8	2.87	3.3	1.3	1.07	1.04	0.68
⅜	8.5	2.48	2.9	1.2	1.07	1.01	0.69
5/16	7.2	2.09	2.5	0.98	1.08	0.99	0.69
¼	5.8	1.69	2.0	0.79	1.09	0.97	0.69
L3 × 3 × ½	9.4	2.75	2.2	1.1	0.90	0.93	0.58
7/16	8.3	2.43	2.0	0.95	0.91	0.91	0.58
⅜	7.2	2.11	1.8	0.83	0.91	0.89	0.59
5/16	6.1	1.78	1.5	0.71	0.92	0.87	0.59
¼	4.9	1.44	1.2	0.58	0.93	0.84	0.59
3/16	3.71	1.09	0.96	0.44	0.94	0.82	0.60
L2½ × 2½ × ½	7.7	2.25	1.2	0.72	0.74	0.81	0.49
⅜	5.9	1.73	0.98	0.57	0.75	0.76	0.49
5/16	5.0	1.47	0.85	0.48	0.76	0.74	0.49
¼	4.1	1.19	0.70	0.39	0.77	0.72	0.49
3/16	3.07	0.90	0.55	0.30	0.78	0.69	0.49

Courtesy of American Institute of Steel Construction

Table 6 Steel angles: unequal legs
Properties for designing; abridged list

Size and thickness	Weight per foot	Area	Axis xx				Axis yy				Axis zz	
in.	lb	in²	I in⁴	I/C in³	r in.	y in.	I in⁴	I/C in³	r in.	x in.	r in.	Tan
L8 × 6 × 1	44.2	13.00	80.8	15.1	2.49	2.65	38.8	8.9	1.73	1.65	1.28	0.543
3/4	33.8	9.94	63.4	11.7	2.53	2.56	30.7	6.9	1.76	1.56	1.29	0.551
1/2	23.0	6.75	44.3	8.0	2.56	2.47	21.7	4.8	1.79	1.47	1.30	0.558
L8 × 4 × 1	37.4	11.00	69.6	14.1	2.52	3.05	11.6	3.9	1.03	1.05	0.85	0.247
3/4	28.7	8.44	54.9	10.9	2.55	2.95	9.4	3.1	1.05	0.95	0.85	0.258
1/2	19.6	5.75	38.5	7.5	2.59	2.86	6.7	2.2	1.08	0.86	0.86	0.267
L6 × 4 × 3/4	23.6	6.94	24.5	6.3	1.88	2.08	8.7	3.0	1.12	1.08	0.86	0.428
1/2	16.2	4.75	17.4	4.3	1.91	1.99	6.3	2.1	1.15	0.99	0.87	0.440
L5 × 3 × 1/2	12.8	3.75	9.5	2.9	1.59	1.75	2.6	1.1	0.83	0.75	0.65	0.357
3/8	9.8	2.86	7.4	2.2	1.61	1.70	2.0	0.89	0.84	0.70	0.65	0.364
1/4	6.6	1.94	5.1	1.5	1.62	1.66	1.4	0.61	0.86	0.66	0.66	0.371
L4 × 3½ × 1/2	11.9	3.50	5.3	1.9	1.23	1.25	3.8	1.5	1.04	1.00	0.72	0.750
3/8	9.1	2.67	4.2	1.5	1.25	1.21	3.0	1.2	1.06	0.96	0.73	0.755
1/4	6.2	1.81	2.9	1.0	1.27	1.16	2.1	0.81	1.07	0.91	0.73	0.759
L4 × 3 × 1/2	11.1	3.25	5.1	1.9	1.25	1.33	2.4	1.1	0.86	0.83	0.64	0.543
3/8	8.5	2.48	4.0	1.5	1.26	1.28	1.9	0.87	0.88	0.78	0.64	0.551
1/4	5.8	1.69	2.8	1.0	1.28	1.24	1.4	0.60	0.90	0.74	0.65	0.558

Size and Thickness		Wt.										
L3½ × 2½ ×	1/2	9.4	2.75	3.2	1.4	1.09	1.20	1.4	0.76	0.70	0.70	0.53	0.486
	7/16	8.3	2.43	2.9	1.3	1.09	1.18	1.2	0.68	0.71	0.68	0.54	0.491
	3/8	7.2	2.11	2.6	1.1	1.10	1.16	1.1	0.59	0.72	0.66	0.54	0.496
	5/16	6.1	1.78	2.2	0.93	1.11	1.14	0.94	0.50	0.73	0.64	0.54	0.501
	1/4	4.9	1.44	1.8	0.75	1.12	1.11	0.78	0.41	0.74	0.61	0.54	0.506
L3 × 2½ ×	1/2	8.5	2.50	2.1	1.0	0.91	1.00	1.3	0.74	0.72	0.75	0.52	0.667
	7/16	7.6	2.21	1.9	0.93	0.92	0.98	1.2	0.66	0.73	0.73	0.52	0.672
	3/8	6.6	1.92	1.7	0.81	0.93	0.96	1.0	0.58	0.74	0.71	0.52	0.676
	5/16	5.6	1.62	1.4	0.69	0.94	0.93	0.90	0.49	0.74	0.68	0.53	0.680
	1/4	4.5	1.31	1.2	0.56	0.95	0.91	0.74	0.40	0.75	0.66	0.53	0.684
L3 × 2 ×	1/2	7.7	2.25	1.9	1.0	0.92	1.08	0.67	0.47	0.55	0.58	0.43	0.414
	7/16	6.8	2.00	1.7	0.89	0.93	1.06	0.61	0.42	0.55	0.56	0.43	0.421
	3/8	5.9	1.73	1.5	0.78	0.94	1.04	0.54	0.37	0.56	0.54	0.43	0.428
	5/16	5.0	1.47	1.3	0.66	0.95	1.02	0.47	0.32	0.57	0.52	0.43	0.435
	1/4	4.1	1.19	1.1	0.54	0.95	0.99	0.39	0.26	0.57	0.49	0.43	0.440
	3/16	3.07	0.90	0.84	0.41	0.97	0.97	0.31	0.20	0.58	0.47	0.44	0.446
L2½ × 2 ×	3/8	5.3	1.55	0.91	0.55	0.77	0.83	0.51	0.36	0.58	0.58	0.42	0.614
	5/16	4.5	1.31	0.79	0.47	0.78	0.81	0.45	0.31	0.58	0.56	0.42	0.620
	1/4	3.62	1.06	0.65	0.38	0.78	0.79	0.37	0.25	0.59	0.54	0.42	0.626
	3/16	2.75	0.81	0.51	0.29	0.79	0.76	0.29	0.20	0.60	0.51	0.43	0.631

Courtesy of American Institute of Steel Construction

Appendix three

Deflections and slopes of beams

Table 1 Deflections and slopes of cantilever beams

y = deflection curve

$y' = \dfrac{dy}{dx}$ = slope of deflection curve

$\delta = y(L)$ = deflection at right end of beam

$\theta = y'(L)$ = slope at right end of beam

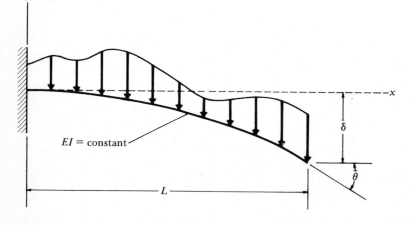

1.

$$y = -\frac{wx^2}{24EI}(6L^2 - 4Lx + x^2)$$

$$y' = -\frac{wx}{6EI}(3L^2 - 3Lx + x^2)$$

$$\delta = -\frac{wL^4}{8EI}, \qquad \theta = -\frac{wL^3}{6EI}$$

Table 1 (continued)

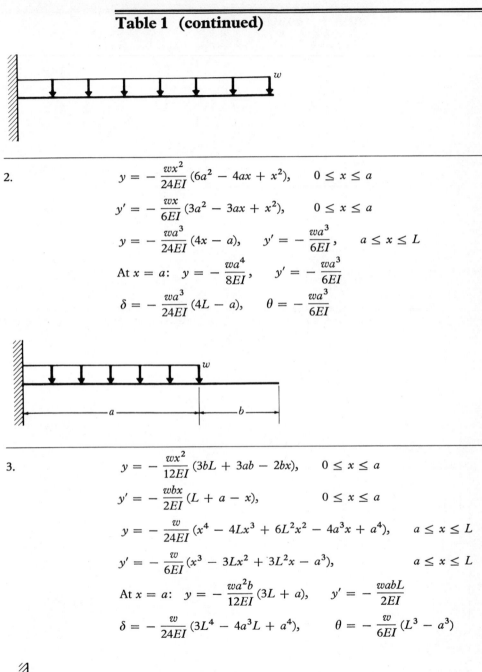

2.

$$y = -\frac{wx^2}{24EI}(6a^2 - 4ax + x^2), \qquad 0 \le x \le a$$

$$y' = -\frac{wx}{6EI}(3a^2 - 3ax + x^2), \qquad 0 \le x \le a$$

$$y = -\frac{wa^3}{24EI}(4x - a), \qquad y' = -\frac{wa^3}{6EI}, \qquad a \le x \le L$$

At $x = a$: $\quad y = -\frac{wa^4}{8EI}, \qquad y' = -\frac{wa^3}{6EI}$

$$\delta = -\frac{wa^3}{24EI}(4L - a), \qquad \theta = -\frac{wa^3}{6EI}$$

3.

$$y = -\frac{wx^2}{12EI}(3bL + 3ab - 2bx), \qquad 0 \le x \le a$$

$$y' = -\frac{wbx}{2EI}(L + a - x), \qquad 0 \le x \le a$$

$$y = -\frac{w}{24EI}(x^4 - 4Lx^3 + 6L^2x^2 - 4a^3x + a^4), \qquad a \le x \le L$$

$$y' = -\frac{w}{6EI}(x^3 - 3Lx^2 + 3L^2x - a^3), \qquad a \le x \le L$$

At $x = a$: $\quad y = -\frac{wa^2b}{12EI}(3L + a), \qquad y' = -\frac{wabL}{2EI}$

$$\delta = -\frac{w}{24EI}(3L^4 - 4a^3L + a^4), \qquad \theta = -\frac{w}{6EI}(L^3 - a^3)$$

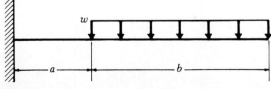

Table 1 (continued)

4.

$$y = -\frac{Px^2}{6EI}(3L - x), \qquad y' = -\frac{Px}{2EI}(2L - x)$$

$$\delta = -\frac{PL^3}{3EI}, \qquad\qquad \theta = -\frac{PL^2}{2EI}$$

5.

$$y = -\frac{Px^2}{6EI}(3a - x), \qquad y' = -\frac{Px}{2EI}(2a - x), \qquad 0 \le x \le a$$

$$y = -\frac{Pa^2}{6EI}(3x - a), \qquad y' = -\frac{Pa^2}{2EI}, \qquad\qquad a \le x \le L$$

At $x = a$: $y = -\dfrac{Pa^3}{3EI}, \qquad y' = -\dfrac{Pa^2}{2EI}$

$$\delta = -\frac{Pa^2}{6EI}(3L - a), \qquad \theta = -\frac{Pa^2}{2EI}$$

6.

$$y = -\frac{M_0 x^2}{2EI}, \qquad y' = -\frac{M_0 x}{EI}$$

$$\delta = -\frac{M_0 L^2}{2EI}, \qquad \theta = -\frac{M_0 L}{EI}$$

Table 1 (continued)

7.
$$y = -\frac{w_0 x^2}{120LEI}(10L^3 - 10L^2 x + 5Lx^2 - x^3)$$

$$y' = -\frac{w_0 x}{24LEI}(4L^3 - 6L^2 x + 4Lx^2 - x^3)$$

$$\delta = -\frac{w_0 L^4}{30EI}, \qquad \theta = -\frac{w_0 L^3}{24EI}$$

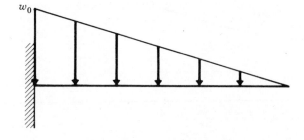

8.
$$y = -\frac{w_0 x^2}{120LEI}(20L^3 - 10L^2 x + x^3)$$

$$y' = -\frac{w_0 x}{24LEI}(8L^3 - 6L^2 x + x^3)$$

$$\delta = -\frac{11w_0 L^4}{120EI}, \qquad \theta = -\frac{w_0 L^3}{8EI}$$

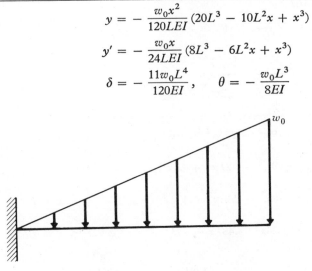

Table 2 Deflections and slopes of simple beams

y = deflection curve

$y' = \dfrac{dy}{dx}$ = slope of deflection curve

$\delta_c = y\left(\dfrac{L}{2}\right)$ = deflection at center of beam

x_1 = distance from A to point of maximum deflection

Table 2 (continued)

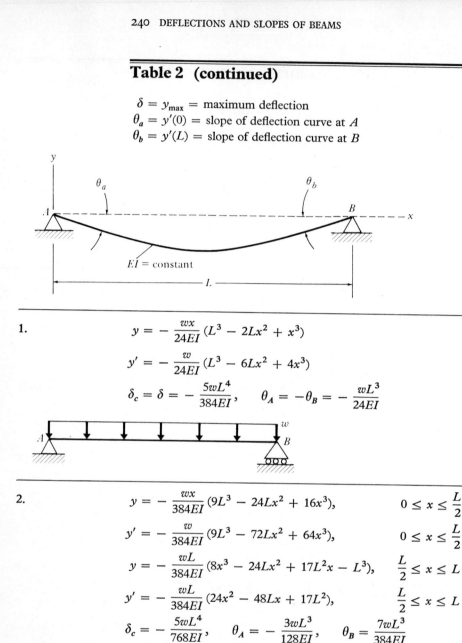

$\delta = y_{max}$ = maximum deflection
$\theta_a = y'(0)$ = slope of deflection curve at A
$\theta_b = y'(L)$ = slope of deflection curve at B

EI = constant

1.
$$y = -\frac{wx}{24EI}(L^3 - 2Lx^2 + x^3)$$

$$y' = -\frac{w}{24EI}(L^3 - 6Lx^2 + 4x^3)$$

$$\delta_c = \delta = -\frac{5wL^4}{384EI}, \qquad \theta_A = -\theta_B = -\frac{wL^3}{24EI}$$

2.
$$y = -\frac{wx}{384EI}(9L^3 - 24Lx^2 + 16x^3), \qquad 0 \le x \le \frac{L}{2}$$

$$y' = -\frac{w}{384EI}(9L^3 - 72Lx^2 + 64x^3), \qquad 0 \le x \le \frac{L}{2}$$

$$y = -\frac{wL}{384EI}(8x^3 - 24Lx^2 + 17L^2x - L^3), \qquad \frac{L}{2} \le x \le L$$

$$y' = -\frac{wL}{384EI}(24x^2 - 48Lx + 17L^2), \qquad \frac{L}{2} \le x \le L$$

$$\delta_c = -\frac{5wL^4}{768EI}, \qquad \theta_A = -\frac{3wL^3}{128EI}, \qquad \theta_B = \frac{7wL^3}{384EI}$$

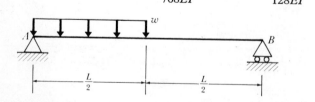

3.
$$y = -\frac{wx}{24LEI}(a^4 - 4a^3L + 4a^2L^2 + 2a^2x^2 - 4aLx^2 + Lx^3),$$
$$0 \le x \le a$$

$$y' = -\frac{w}{24LEI}(a^4 - 4a^3L + 4a^2L^2 + 6a^2x^2 - 12aLx^2 + 4Lx^3),$$
$$0 \le x \le a$$

Table 2 (continued)

$$y = -\frac{wa^2}{24LEI}(-a^2L + 4L^2x + a^2x - 6Lx^2 + 2x^3), \qquad a \le x \le L$$

$$y' = -\frac{wa^2}{24LEI}(4L^2 + a^2 - 12Lx + 6x^2), \qquad a \le x \le L$$

$$\theta_A = -\frac{wa^2}{24LEI}(a^2 - 4aL + 4L^2), \qquad \theta_B = \frac{wa^2}{24LEI}(2L^2 - a^2)$$

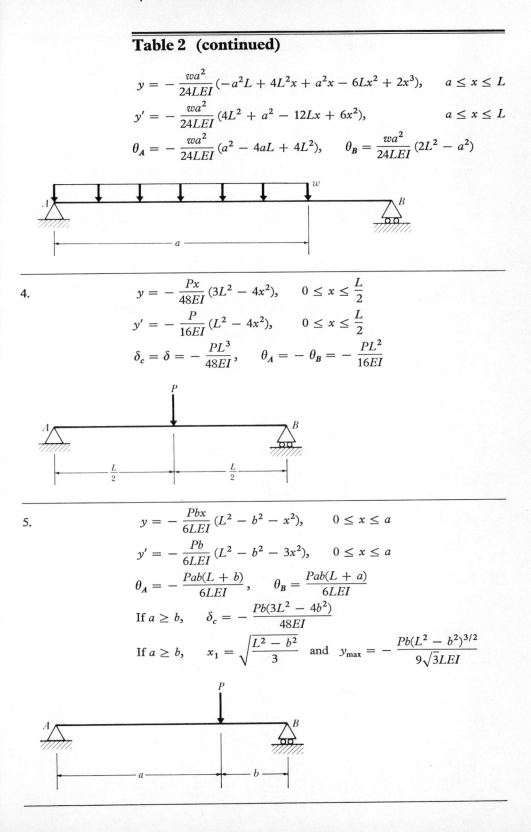

4.

$$y = -\frac{Px}{48EI}(3L^2 - 4x^2), \qquad 0 \le x \le \frac{L}{2}$$

$$y' = -\frac{P}{16EI}(L^2 - 4x^2), \qquad 0 \le x \le \frac{L}{2}$$

$$\delta_c = \delta = -\frac{PL^3}{48EI}, \qquad \theta_A = -\theta_B = -\frac{PL^2}{16EI}$$

5.

$$y = -\frac{Pbx}{6LEI}(L^2 - b^2 - x^2), \qquad 0 \le x \le a$$

$$y' = -\frac{Pb}{6LEI}(L^2 - b^2 - 3x^2), \qquad 0 \le x \le a$$

$$\theta_A = -\frac{Pab(L + b)}{6LEI}, \qquad \theta_B = \frac{Pab(L + a)}{6LEI}$$

If $a \ge b$, $\qquad \delta_c = -\dfrac{Pb(3L^2 - 4b^2)}{48EI}$

If $a \ge b$, $\qquad x_1 = \sqrt{\dfrac{L^2 - b^2}{3}}$ and $y_{max} = -\dfrac{Pb(L^2 - b^2)^{3/2}}{9\sqrt{3}LEI}$

Table 2 (continued)

6.

$$y = -\frac{Px}{6EI}(3aL - 3a^2 - x^2), \qquad 0 \le x \le a$$

$$y' = -\frac{P}{2EI}(aL - a^2 - x^2), \qquad 0 \le x \le a$$

$$y = -\frac{Pa}{6EI}(3Lx - 3x^2 - a^2), \qquad a \le x \le \frac{L}{2}$$

$$y' = -\frac{Pa}{2EI}(L - 2x), \qquad a \le x \le \frac{L}{2}$$

$$\theta_A = -\theta_B = -\frac{Pa(L-a)}{2EI}, \qquad \delta_c = y_{max} = -\frac{Pa}{24EI}(3L^2 - 4a^2)$$

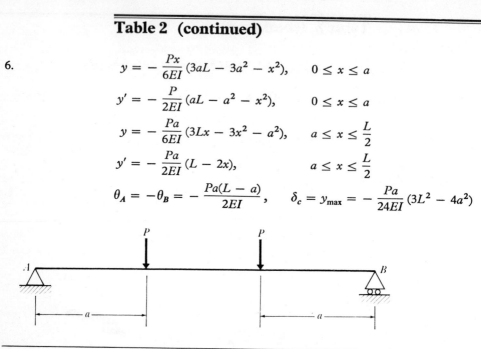

7.

$$y = -\frac{M_0 x}{6LEI}(2L^2 - 3Lx + x^2)$$

$$y' = -\frac{M_0}{6LEI}(2L^2 - 6Lx + 3x^2)$$

$$\delta_c = -\frac{M_0 L^2}{16EI}, \qquad \theta_A = -\frac{M_0 L}{3EI}, \qquad \theta_B = \frac{M_0 L}{6EI}$$

$$x_1 = L\left(1 - \frac{\sqrt{3}}{3}\right) \quad \text{and} \quad \delta_{max} = -\frac{M_0 L^2}{9\sqrt{3}EI}$$

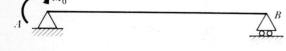

8.

$$y = -\frac{M_0 x}{24LEI}(L^2 - 4x^2), \qquad 0 \le x \le \frac{L}{2}$$

$$y' = -\frac{M_0}{24LEI}(L^2 - 12x^2), \qquad 0 \le x \le \frac{L}{2}$$

$$\delta_c = 0, \qquad \theta_A = -\frac{M_0 L}{24EI}, \qquad \theta_B = -\frac{M_0 L}{24EI}$$

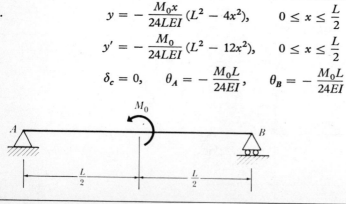

Table 2 (continued)

9.

$$y = -\frac{M_0 x}{6LEI}(6aL - 3a^2 - 2L^2 - x^2), \qquad 0 \le x \le a$$

$$y' = -\frac{M_0}{6LEI}(6aL - 3a^2 - 2L^2 - 3x^2), \qquad 0 \le x \le a$$

At $x = a$: $y = -\dfrac{M_0 a}{3LEI}(3aL - 2a^2 - L^2)$

At $x = a$: $y' = -\dfrac{M_0}{3LEI}(3aL - 3a^2 - L^2)$

$$\theta_A = -\frac{M_0}{6LEI}(6aL - 3a^2 - 2L^2), \qquad \theta_B = \frac{M_0}{6LEI}(3a^2 - L^2)$$

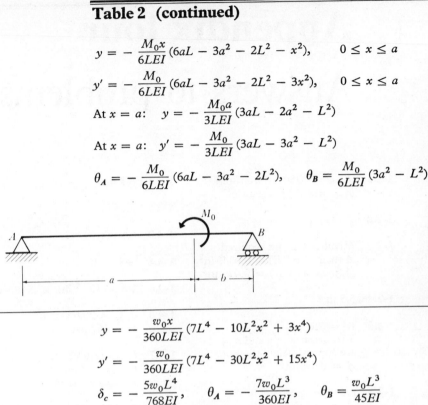

10.

$$y = -\frac{w_0 x}{360LEI}(7L^4 - 10L^2 x^2 + 3x^4)$$

$$y' = -\frac{w_0}{360LEI}(7L^4 - 30L^2 x^2 + 15x^4)$$

$$\delta_c = -\frac{5w_0 L^4}{768EI}, \qquad \theta_A = -\frac{7w_0 L^3}{360EI}, \qquad \theta_B = \frac{w_0 L^3}{45EI}$$

$$x_1 = 0.5193L, \qquad \delta = -0.00652\frac{w_0 L^4}{EI}$$

Appendix four

Answers to problems

1-2.1 $\sigma_{\text{cable}} = 6350$ psi (tension)
$\sigma_{\text{link}} = 8000$ psi (tension)

1-2.2 10.4 mm

1-2.3 Member CB not required
A (member DE) = 0.0345 in^2
A (member CE) = 0.077 in^2

1-2.4 5.5 m

1-2.5 $W = 2900$ lb
Yes, because σ_{BC} and σ_{AB}
are approximately the same
$A = 0.099$ in^2

1-2.6 $z = 2$ ft, $x = 3$ ft
$A_B = 0.033$ in^2
$A_C = 0.05$ in^2

1-3.1 140,000 lb

1-3.2 $-\frac{2}{9} \times 10^{-3}$ in. for steel
$\frac{5}{8} \times 10^{-3}$ in. for aluminum

1-3.3 0.043 mm

1-3.4 $x = 4$ ft
$A = 100$ in^2
$a = 10$ in.

1-3.5 $D = 9.8$ mm
$d = 24$ mm,
$s = 120$ mm

1-3.6 $\epsilon_{AD} = -7$ mm/m
$\epsilon_{BC} = 4$ mm/m

1-3.7 $\tau_A = 2650$ psi
$\tau_E = 1920$ psi
$\tau_H = 5300$ psi
$\tau_D = 5370$ psi
$A = 0.06$ in^2

1-4.1 6.5 mm

1-4.2 0.25 in.

1-4.3 0.6 mm

1-4.4 -3×10^{-6} m

1-4.5 1.5×10^{-4} in.

1-4.6 $\sigma_{\text{ring}} = 20,000$ psi
$P = 167$ psi

1-4.7 $t_{\text{sphere}} = 0.07$ in.
$t_{\text{cylinder}} = 0.07$ in.
$L_{\text{cylinder}} = 34$ in.

2-1.1 $\sigma = 375$ psi (compression)
$\tau = 217$ psi

2-1.2 $\sigma = 4.1$ MPa (tension)
$\tau = 17$ MPa

2-1.3 $\sigma = 27.7$ MPa (compression)
$\tau = 4.2$ MPa

2-1.4 $\sigma = 1200$ psi (compression)
$\tau = 0$

2-1.5 $\sigma_x = 100$ psi (tension)
$\sigma_y = 31$ psi (tension)

2-1.6 $\tau_{xy} = \tau_{yx} = -323$ MPa
$\sigma_x = 164$ MPa (tension)

2-1.7 $\sigma_{\text{max}} = \sigma_x$
$\sigma_{\text{min}} = \sigma_{\text{at } \pi/2} = 0$
$\tau_{\text{max}} = \tau_{\text{min}} = \pm\sigma_{x/2}$

2-1.8 $P \leq 50.7$ kN

2-2.5 $\sigma_\theta = 16.2$ MPa (compression)
$\tau_\theta = 0$

2-2.6 For $\theta = 120°$:
$\sigma_\theta = 40$ ksi (tension)
$\tau_\theta = 0$
For $\theta = 30°$:
$\sigma_\theta = 0$, $\tau_\theta = 0$

2-3.1 $\sigma_{p1} = 15$ ksi,
$\sigma_{p2} = -35$ ksi
$\tau_{max} = 25$ ksi

2-3.2 $\sigma_{p1} = 290$ MPa,
$\sigma_{p2} = 30$ MPa
$\tau_{max} = 130$ MPa

2-3.3 $\sigma_{p1} = -1000$ psi,
$\sigma_{p2} = -3000$ psi
$\tau_{max} = 1000$ psi

2-3.4 $\sigma_{p1} = 169$ MPa,
$\sigma_{p2} = -29$ MPa
$\tau_{max} = 99$ MPa

2-3.5 $\sigma_{p1} = 8000$ psi,
$\sigma_{p2} = -2000$ psi
$\tau_{max} = 5000$ psi

2-3.6 $\sigma_{p1} = -5$ MPa,
$\sigma_{p2} = -25$ MPa
$\tau_{max} = 10$ MPa

2-3.7 $\sigma_{p1} = 9.1$ ksi,
$\sigma_{p2} = -7.1$ ksi
$\tau_{max} = 8.1$ ksi

2-3.8 $\sigma_{p1} = 290$ MPa,
$\sigma_{p2} = -210$ MPa
$\tau_{max} = 250$ MPa

2-4.5 (a) $\sigma_{p1} = -300$ MPa,
$\sigma_{p2} = -300$ MPa
$\tau_{max} = 0$
(b) $\sigma_{p1} = 40$ MPa, $\sigma_{p2} = 0$
$\tau_{max} = 20$ MPa
(c) $\sigma_{p1} = 20$ ksi,
$\sigma_{p2} = -20$ ksi
$\tau_{max} = 20$ ksi

2-4.6 $\sigma_{45°} = 92.5$ MPa
$\sigma_{-45°} = 22.5$ MPa
$\tau_{45°} = 82.5$ MPa

2-5.1 $|\tau_{max}| = 25$ ksi

2-5.2 $|\tau_{max}| = 145$ MPa

2-5.3 $|\tau_{max}| = 1500$ psi

2-5.4 $|\tau_{max}| = 99$ MPa

2-5.5 150 psi, 0 psi

2-5.6 $|\tau_{max}| = 89.6$ MPa

2-5.7 $|\tau_{max}| = 17$ MPa

2-5.8 $|\tau_{max}| = 35$ MPa

2-6.2 3×10^{-3} m/m, 336 kN

2-6.3 $\epsilon_d = -5 \times 10^{-3}$ m/m
$\epsilon_e = 15 \times 10^{-3}$ m/m

2-6.4 2×10^{-4} in./in.

2-6.5 $60°$

2-6.6 $\epsilon_{45°} = 300$ μm/m

2-6.7 $\epsilon_u = 5.87 \times 10^{-3}$ in./in.
$\epsilon_v = 2.13 \times 10^{-3}$ in./in.
$\gamma_{uv} = 2.46 \times 10^{-3}$ rad

2-6.8 $\theta = -18°$ from u axis
$\epsilon_u = -930$ μm/m
$\epsilon_v = 430$ μm/m

2-7.1 $\epsilon_{p1} = 2300$ μin./in.,
$\epsilon_{p2} = -300$ μin./in.
$\gamma_{max} = 2600$ μrad

2-7.2 $\epsilon_{p1} = 1000$ μin./in.,
$\epsilon_{p2} = -1000$ μin./in.
$\gamma_{max} = 2000$ μrad

2-7.3 $\epsilon_{p1} = 1200$ μin./in.,
$\epsilon_{p2} = -800$ μin./in.
$\gamma_{max} = 2000$ μrad

2-7.4 $\epsilon_{p1} = 61.4 \times 10^{-3}$ m/m,
$\epsilon_{p2} = -41.4 \times 10^{-3}$ m/m
$\gamma_{max} = 103 \times 10^{-3}$ rad

2-7.5 $\epsilon_{p1} = -5.8 \times 10^{-3}$ in./in.,
$\epsilon_{p2} = -34.2 \times 10^{-3}$ in./in.
$\gamma_{max} = 28.4 \times 10^{-3}$ rad

2-7.6 $\epsilon_{p1} = 180$ μm/m,
$\epsilon_{p2} = -320$ μm/m
$\gamma_{max} = 500$ μrad

2-7.7 $\epsilon_x = 466$ μin./in.
$\epsilon_y = -1266$ μin./in.
$\gamma_{xy} = 1000$ μrad

2-8.1 $\epsilon_x = 2500\ \mu\text{in./in.}$,
$\epsilon_y = -500\ \mu\text{in./in.}$
$\gamma_{xy} = -5200\ \mu\text{rad}$

2-8.2 $\epsilon_x = 2000\ \mu\text{in./in.}$,
$\epsilon_y = -3000\ \mu\text{in./in.}$
$\gamma_{xy} = 0$
$\epsilon_{p1} = 2000\ \mu\text{in./in.}$,
$\epsilon_{p2} = -3000\ \mu\text{in./in.}$
$\gamma_{\text{max}} = 5000\ \mu\text{rad}$

2-8.3 $\epsilon_{p1} = 25 \times 10^{-3}\ \text{m/m}$,
$\epsilon_{p2} = -15 \times 10^{-3}\ \text{m/m}$

2-8.4 $\epsilon_{p1} = \epsilon_y, \epsilon_{p2} = \epsilon_x$
$\sigma_{p1} = 180{,}000\ \text{psi}, \sigma_{p2} = 0$

2-8.5 $\epsilon_{p1} = 1000\ \mu\text{m/m}$,
$\epsilon_{p2} = 200\ \mu\text{m/m}$

2-8.6 $\epsilon_{p1} = 234\ \mu\text{m/m}$,
$\epsilon_{p2} = -264\ \mu\text{m/m}$

2-8.7 $\epsilon_{p1} = 302\ \mu\text{m/m}$,
$\epsilon_{p2} = -402\ \mu\text{m/m}$

2-9.1 $\sigma_x = 76{,}000\ \text{psi}$
$\sigma_y = 4000\ \text{psi}$
$\tau_{xy} = -62{,}400\ \text{psi}$
$\sigma_{p1} = 112{,}000\ \text{psi}$
$\sigma_{p2} = -32{,}000\ \text{psi}$

2-9.3 $\sigma_x = -240\ \text{MPa}$,
$\sigma_y = 80\ \text{MPa}$
$\tau_{xy} = -120\ \text{MPa}$
$\sigma_{p1} = 120\ \text{MPa}$,
$\sigma_{p2} = -280\ \text{MPa}$

2-9.4 $\epsilon_{p1} = 4800\ \mu\text{m/m}$,
$\epsilon_{p2} = -5600\ \mu\text{m/m}$
$\sigma_{p1} = 231\ \text{MPa}$,
$\sigma_{p2} = -315\ \text{MPa}$

2-9.5 $\epsilon_{p1} = 606\ \mu\text{in./in.}$,
$\epsilon_{p2} = 394\ \mu\text{in./in.}$
$\sigma_{p1} = 22{,}500\ \text{psi}$,
$\sigma_{p2} = 17{,}500\ \text{psi}$

2-9.6 $\sigma_x = 24\ \text{MPa}$,
$\sigma_y = -16\ \text{MPa}$
$\tau_{xy} = -48\ \text{MPa}$
$\sigma_{p1} = 56\ \text{MPa}$,
$\sigma_{p2} = -48\ \text{MPa}$

2-9.7 $\epsilon_{p1} = 4800\ \mu\text{m/m}$,
$\epsilon_{p2} = -5600\ \mu\text{m/m}$
$\sigma_{p1} = 231\ \text{MPa}$,
$\sigma_{p2} = -315\ \text{MPa}$

2-9.8   320 psi

3-1.1   15,900 psi between B and C

3-1.2   2.89

3-1.3 $d_i = 56\ \text{mm}$
$\tau_{\text{solid}} = 71.3\ \text{MPa}$
$\tau_{\text{o (hollow)}} = 30.7\ \text{MPa}$
$\tau_{\text{i (hollow)}} = 22.9\ \text{MPa}$

3-1.4   8530 psi

3-1.5   25%

3-1.6   119 MPa in the steel

3-1.7   56,500 lb-in.
3.6 in.

3-1.8   6.9 MPa in section AB

3-2.1   157 in.
1.57 lb-in.

3-2.2 $3.18 \times 10^{-3}\ \text{rad}$

3-2.3   0.024 rad

3-2.4   0.058 rad

3-2.5 $2.19 \times 10^{-3}\ \text{rad}$

3-2.6   85 mm

3-2.7   6 in.

3-2.8   57 mm

3-2.9 $TL/2JG$

3-3.1   13,400 psi

3-3.2   52 mm
1740 N·m
$\sigma = 60.5\ \text{MPa}$,
$\tau = 17.6\ \text{MPa}$

3-3.3   1.55 in.

3-3.4 $24.3 \times 10^3\ \text{N·m}$
$\sigma = 73\ \text{MPa}$,
$\tau = 42\ \text{MPa}$

3-3.5 $0.36 \times 10^{-3}\ \text{m/m}$
(compression)

3-3.6   550 $\mu\text{in./in.}$ (compression)

3-3.7   6530 lb-in.

3-4.1   1.04 in.

3-4.2   49 mm

3-4.3   155 MPa

3-4.4   16,000 psi

3-4.5   135 hp

3-4.6   90 mm

3-4.7   4.1 in.

3-4.8   115 mm

4-1.1

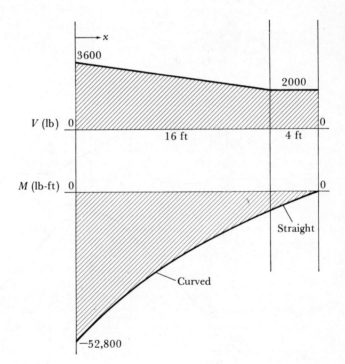

$$V_{0 \to 16} = 3600 - 100x$$
$$V_{16 \to 20} = 3600 - 1600$$
$$M_{0 \to 16} = -52{,}800 + 3600x$$
$$- 100x \left(\frac{x}{2} \right)$$
$$M_{16 \to 20} = -52{,}800$$
$$+ 3600x$$
$$- 1600(x - 8)$$

4-1.2

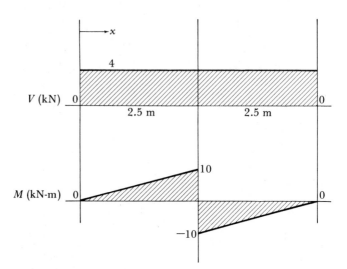

$$V_{0\to5} = 4$$
$$M_{0\to2.5} = 4x$$
$$M_{2.5\to5} = 4x - 20$$

4-1.3

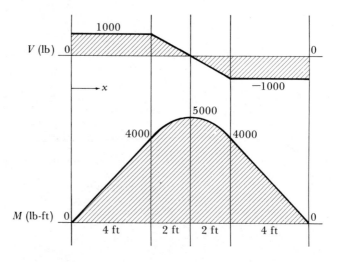

$$V_{0\to4} = 1000$$
$$V_{4\to8} = 1000 - 500(x - 4)$$
$$V_{8\to12} = 1000 - 2000$$
$$M_{0\to4} = 1000x$$
$$M_{4\to8} = 1000x$$
$$\quad\quad - 500(x - 4)$$
$$\quad\quad \times \frac{x - 4}{2}$$
$$M_{8\to12} = 1000x$$
$$\quad\quad - 2000(x - 6)$$

4-1.4

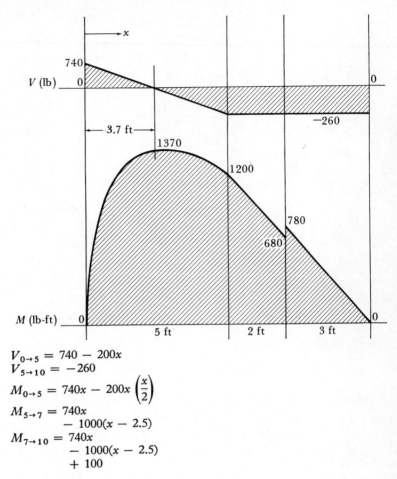

$V_{0 \to 5} = 740 - 200x$

$V_{5 \to 10} = -260$

$M_{0 \to 5} = 740x - 200x \left(\dfrac{x}{2}\right)$

$M_{5 \to 7} = 740x$
$\qquad - 1000(x - 2.5)$

$M_{7 \to 10} = 740x$
$\qquad - 1000(x - 2.5)$
$\qquad + 100$

4-1.5

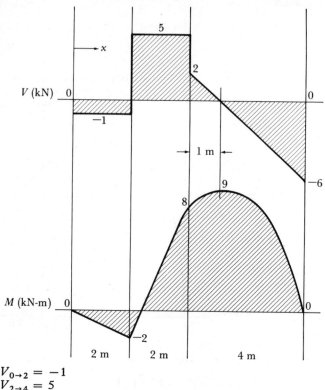

$V_{0 \to 2} = -1$
$V_{2 \to 4} = 5$
$V_{4 \to 8} = 10 - 2x$
$M_{0 \to 2} = -x$
$M_{2 \to 4} = 5x - 12$
$M_{4 \to 8} = 2x - (x - 4)^2$

4-1.6

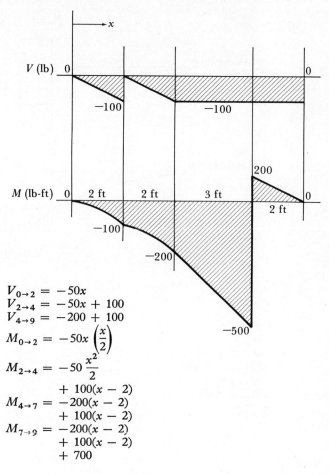

$$V_{0 \to 2} = -50x$$
$$V_{2 \to 4} = -50x + 100$$
$$V_{4 \to 9} = -200 + 100$$
$$M_{0 \to 2} = -50x \left(\frac{x}{2}\right)$$
$$M_{2 \to 4} = -50 \frac{x^2}{2}$$
$$\qquad\quad + 100(x - 2)$$
$$M_{4 \to 7} = -200(x - 2)$$
$$\qquad\quad + 100(x - 2)$$
$$M_{7 \to 9} = -200(x - 2)$$
$$\qquad\quad + 100(x - 2)$$
$$\qquad\quad + 700$$

4-1.7

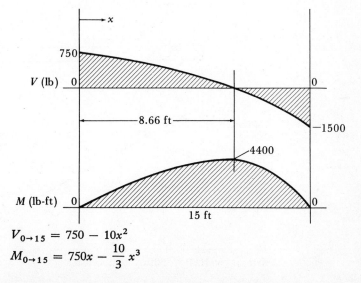

$$V_{0 \to 15} = 750 - 10x^2$$
$$M_{0 \to 15} = 750x - \frac{10}{3} x^3$$

4-1.8

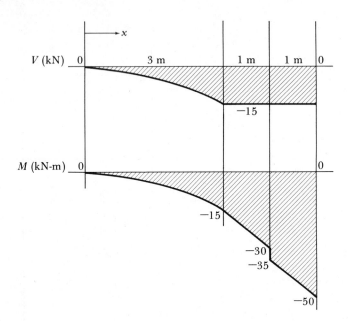

$V_{0\to3} = -\frac{5}{3}x^2$
$V_{3\to5} = -15$
$M_{0\to3} = -\frac{5}{9}x^3$
$M_{3\to4} = -30 - 15x$
$M_{4\to5} = 25 - 15x$

4-1.9

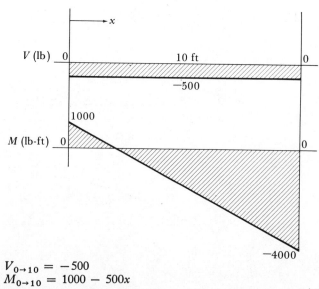

$V_{0\to10} = -500$
$M_{0\to10} = 1000 - 500x$

4-1.10

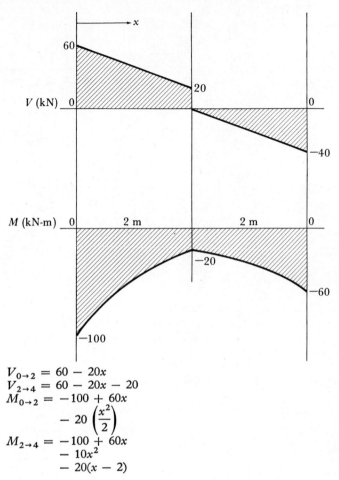

$$V_{0 \to 2} = 60 - 20x$$
$$V_{2 \to 4} = 60 - 20x - 20$$
$$M_{0 \to 2} = -100 + 60x$$
$$- 20 \left(\frac{x^2}{2} \right)$$
$$M_{2 \to 4} = -100 + 60x$$
$$- 10x^2$$
$$- 20(x - 2)$$

4-1.11

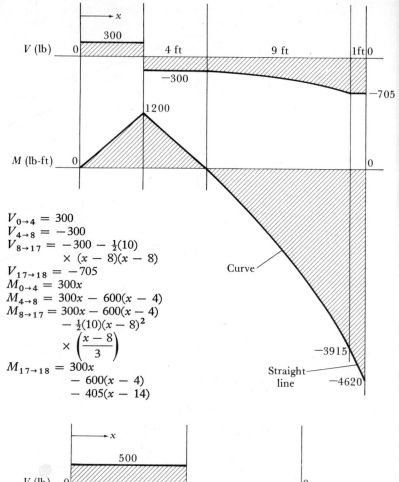

$V_{0\to4} = 300$
$V_{4\to8} = -300$
$V_{8\to17} = -300 - \frac{1}{2}(10)$
$\qquad \times (x-8)(x-8)$
$V_{17\to18} = -705$
$M_{0\to4} = 300x$
$M_{4\to8} = 300x - 600(x-4)$
$M_{8\to17} = 300x - 600(x-4)$
$\qquad - \frac{1}{2}(10)(x-8)^2$
$\qquad \times \left(\dfrac{x-8}{3}\right)$
$M_{17\to18} = 300x$
$\qquad - 600(x-4)$
$\qquad - 405(x-14)$

4-1.12

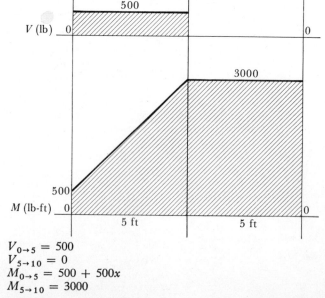

$V_{0\to5} = 500$
$V_{5\to10} = 0$
$M_{0\to5} = 500 + 500x$
$M_{5\to10} = 3000$

4-2.1 513 psi (compression)

4-2.2 34.7 mm

4-2.3 $\sigma_{\text{top}} = 6750$ psi (tension)
$\sigma_{\text{bottom}} = 13,500$ psi
(compression)

4-2.4 (a) 760 kPa
(b) 26.7% increase in stress
62.5% decrease in mass

4-2.5 237 in.

4-2.6 14.4 MPa maximum tensile
stress
29.8 MPa maximum comp.
stress

4-2.7 3560 lb

4-2.8 5.95 kN

4-3.1 193 kg/m

4-3.3 7.86 kN

4-3.4 2 in./nail

4-3.5 4.2 in. between rivets along
length of beam

4-3.6 884 kPa, 25 mm from top
1.59 MPa, 50 mm from top
2.12 MPa, 75 mm from top
2.48 MPa, 100 mm from top
2.65 MPa, 125 mm from top
2.65 MPa, 150 mm from top
2.48 MPa, 175 mm from top
2.12 MPa, just above junction

4-3.7 1810 psi

4-4.1 At A: $\sigma_{p1} = 0$,
$\sigma_{p2} = -2220$ psi
At B: $\sigma_{p1} = 38$ psi,
$\sigma_{p2} = -1118$ psi
At C: $\sigma_{p1} = -1960$ psi,
$\sigma_{p2} = 0$

4-4.2 At A: $\sigma_{p1} = -1.8$ MPa,
$\sigma_{p2} = -120$ MPa
At B: $\sigma_{p1} = 0$,
$\sigma_{p2} = -117$ MPa
At C: $\sigma_{p1} = 4.5$ MPa,
$\sigma_{p2} = -4.5$ MPa
At D: $\sigma_{p1} = 60.1$ MPa,
$\sigma_{p2} = -0.09$ MPa

4-4.3 393 kN

4-4.4 Top: $\sigma_{p1} = 13,900$ psi,
$\sigma_{p2} = 0$
$\tau_{\text{max}} = 6970$ psi
Junction: $\sigma_{p1} = 15,400$ psi,
$\sigma_{p2} = -2520$ psi
$\tau_{\text{max}} = 8960$ psi
n.a.: $\sigma_{p1} = 7740$ psi,
$\sigma_{p2} = -7740$ psi
$\tau_{\text{max}} = 7740$ psi

5-1.1 $\dfrac{Ml}{EI}, \dfrac{Ml^2}{2EI}$

5-1.2 $\dfrac{7.03 \times 10^3}{EI}, \dfrac{4.89}{EI}$ km

5-1.3 0.56ℓ in. from left end

5-1.4 $-\dfrac{10}{EI}$ km midspan

5-1.5 $-\dfrac{11p_0\ell^4}{120EI}$ in., $-\dfrac{p_0\ell^3}{8EI}$

5-1.6 $EIy = 334x^2 - 66.7x^3$
$6.25x^4 - 0.21x^5$

5-1.7 $y_A = -\dfrac{9.34 \times 10^6}{EI}$ in.
$y_B = \dfrac{32 \times 10^6}{EI}$ in.

5-1.8 $-\dfrac{5.75 \times 10^3}{EI}$

5-1.9 $-\dfrac{83.2L^4}{EI}$ in.

5-1.10 $-\dfrac{0.117L^4w}{EI}$ in.

5-1.11 $-\dfrac{11.5}{EI}$ km, $-\dfrac{8.17 \times 10^3}{EI}$

5-1.12 $\dfrac{0.00293L^3P}{EI}$ in.

5-1.13 $-\dfrac{0.0274PL^3}{EI}$ in.

5-1.14 $\dfrac{9.78 \times 10^3}{EI}$

5-1.15 $-\dfrac{3.67}{EI}$ km

5-2.1 $EIy = \frac{850}{3}x^3 - \frac{125}{6}x^4$
$+ \frac{125}{6}\langle x - 6\rangle^4$
$- 500\langle x - 8\rangle^3$
$+ \frac{4150}{3}\langle x - 10\rangle^3$
$- \frac{100}{3}\langle x - 10\rangle^4$
$- 7630x$

5-2.2 $-\dfrac{139{,}000}{EI}$ ft

5-2.3 650 N

5-2.4 $-\dfrac{2200}{EI}$

5-2.5 $-\dfrac{0.00111p_0L^4}{EI}$ in.

5-2.6 $-\dfrac{21.5}{EI}$ km

5-2.7 $-\dfrac{319{,}000}{EI}$ ft

6-1.1 $\sigma_A = 750$ psi
$\sigma_B = -1250$ psi
$\sigma_C = -350$ psi
$\sigma_D = 1650$ psi

6-1.2 46.8 kN

6-1.3 $\sigma_{p1} = 30.7$ MPa
$\sigma_{p2} = -18.2$ MPa
$\sigma_{p3} = 0$
$|\tau_{max}| = 24.5$ MPa

6-1.4

$\sigma = 8270$ psi
$\tau = 1570$ psi

6-2.1 11.4 kN

6-2.2 $\sigma_{p1} = 680$ psi
$\sigma_{p2} = -4930$ psi
$\sigma_{p3} = 0$
$|\tau_{max}| = 2800$ psi

6-2.3 Just to the right of gear A,
and on top of the shaft,
$\sigma_{p1} = 7945$ psi
$\sigma_{p2} = -1835$ psi
$|\tau_{max}| = 4890$ psi

6-2.4 $\sigma_y = 109$ MPa

$\sigma_x = 54.5$ MPa
$\tau_{xy} = 13.2$ MPa

$\sigma_{p1} = 112$ MPa
$\sigma_{p2} = 51.5$ MPa
$\sigma_{p3} = 0$
$|\tau_{max}| = 56$ MPa

7-1.1 $\sigma_s = 8570$ psi
$\sigma_c = 714$ psi

7-1.2 $F_A = 39.8$ kN
$F_B = 10.2$ kN

7-1.3 $P_S = 2.62$ lb, $P_A = 3.69$ lb

7-1.4 $\sigma_S = 9730$ psi, $\sigma_B = 8420$ psi

7-1.5 $P_B = 20.3$ kN (compression)
$P_S = 20.3$ kN (tension)

7-1.6 $F_B = 3$ kN, $F_C = 6$ kN

7-1.7 54,300 lb

7-2.1 242,000 lb-in.

7-2.2 402,000 lb-in.

7-2.3 $\tau_S = 7170$ psi, $\tau_B = 1430$ psi
$\theta = 0.0186$ rad

7-2-4 84,000 lb-in.

7-2.5 6.52 MPa

7-2.6 $T_A = 133$ N·m
$T_B = 633$ N·m

7-2.7 0.0206 rad

7-3.1 $F_A = 1.5$ kN
$F_B = 2.5$ kN, $M_B = 2$ kN·m

7-3.2 $F_A = 3.72$ kN
$F_B = -427$ N
$F_C = 1.85$ kN

7-3.3 $0.141 \, wL$

7-3.4 $\dfrac{5wL^4}{8\gamma AL^3 + 384EI}$

7-3.5 $F_B = 2.93p_0$ N,
$M_B = 2.89p_0$ N·m
$F_A = 1.57p_0$ N,
$M_A = 2.19p_0$ N·m

7-3.6 $0.369wL$

7-4.1 $\delta = \dfrac{5PL^3}{2KL^3 + 48EI}$

7-4.2 $F_B = 337$ N, $F_A = 263$ N
$T_A = 135$ N·m,
$M_A = 263$ N·m

7-4.3 $\sigma_{max} = 11,200$ psi (tension on top and compression on bottom)
$\tau_{max} = 3,200$ psi

7-4.4 $F_A = F_B = \dfrac{P}{2}$
$M_A = M_B = 11.3P$ lb-in.

8-3.1 1.24

8-3.2 35.4%

8-3.3 $\theta = 0.18$ rad

8-3.4 31 mm, 0.15 rad

8-3.5 325,000 lb-in.

8-3.6 (a) 5.13×10^{-3} rad
(b) 4.970 kN·m

8-3.7 $T_e = 103,000$ lb-in.
$T_{FP} = 116,000$ lb-in.

8-3.8 $T_e = 25.8$ kN·m
$T_{FP} = 32.1$ kN·m

8-4.1 1.7

8-4.2 1.5

8-4.3 1.76

8-4.4 1.23

8-4.5 $M_e = 138.1\sigma_e$
$M_{FP} = 149.5\sigma_e$

8-4.6 $M_e = 156 \times 10^{-6}\sigma_e$ N·m
$M_{FP} = 219 \times 10^{-6}\sigma_e$ N·m

8-4.7 $P = 80,000$ lb
$0 \le x \le 20.4$ in.,
51.6 in. $\le x \le 72$ in.
where x is measured from left end

8-4.8 $M_{FP} = 341$ kN·m
$M_e = 199$ kN·m

9-1.1 0.879 in.

9-1.2 1.2 mm (down)
0.232 mm (right)

9-1.3 (a) $\dfrac{2w\ell^3}{3EI}$

(b) $y = \dfrac{11w\ell^4}{24EI} + \dfrac{9w\ell^2}{5GA}$

$\theta = \dfrac{2w\ell^3}{3EI}$

(c) $6.57\sqrt{E/G}$

9-1.4 $\dfrac{T_0\ell}{GJ} + \dfrac{T_0 b}{EI}$

9-1.5 $\dfrac{Pab^2}{8EI_b}$ (right)

9-1.6 0.0624 in. (right)
0.061 in. (down)

9-1.7 $\dfrac{RT_0\pi}{2}\left(\dfrac{1}{EI} + \dfrac{1}{GJ}\right)$

9-2.1 $P_{rod} = 6.27$ kN
$V_A = 25$ kN,
$M_A = 22.5$ kN·m

9-2.3 Simply supported end:
$V = 1.39$ kN
Clamped end:
$T = 834$ N·m
$V = 6.61$ kN
$M = 3.97$ kN·m

9-2.4 $y_C = \dfrac{P\ell_B^3}{3E_B I_B}$

$P_C = \dfrac{3w\ell_A^4 E_B I_B}{8(\ell_A^3 E_B I_B + \ell_B^3 E_A I_A)}$

9-2.5 $V_A = V_B$

$= \dfrac{P\ell^2}{16hI_0\left(\dfrac{h}{3I_1} + \dfrac{\ell}{2I_0}\right)}$

$P_A = P_B = \dfrac{P}{2}$

9-2.6 $T_A = T_C$

$$= \frac{Pb^2}{4EI\left(\dfrac{a}{GJ} + \dfrac{b}{EI}\right)}$$

10-1.1 326 in.

10-1.2 12.7°F

10-1.3 52 kN

10-1.4 36.5 mm

10-1.5 924 N

10-1.6 554 lb

10-2.1 34,300 lb

10-2.2 319,000 lb

10-2.3 $P = 37,500$ lb
$\sigma_{max} = -16,300$ psi

10-2.4 156 kN

10-2.5 10.3×10^3 kg

10-3.1 26,000 lb

10-3.2 48,400 psi

10-3.3 2.65 MN